"The AI Shift *brilliantly captures the moment we are living in. This moment demands awareness, adaptability, and action. With contributions from a diverse group of global experts, this book transforms a complex technological revolution into something practical and empowering. It doesn't just explain AI; it shows you how to engage with it. If you want to stay relevant, productive, and purposeful in the years ahead, this is essential reading."*

—MELISSA GEARD, AI Strategy Advisor

"The co-authors of this book have created a timely and accessible guide to navigating one of the most profound transformations in human history. The AI Shift *removes fear and replaces it with clarity, offering real-world insights on how to thrive alongside intelligent machines. This is not a book about technology; it's a book about opportunity, mindset, and the future we choose to build."*

—JAMES HANLEY, CEO of AIworks

"In a world where AI can feel overwhelming, The AI Shift *provides exactly what we need: perspective. Drawing from voices across industries and continents, this book offers a grounded, optimistic path forward. It reminds us that the future isn't happening to us. It's something we actively shape. This book is powerful resource for anyone looking to turn uncertainty into advantage."*

—PRIYA NAINiSH, Global Head of Digital Transformation

"The AI Shift *is a wake-up call delivered with clarity and direction. Much like the late-night call described in its introduction, this book challenges readers to respond thoughtfully to a rapidly changing world. It equips readers with the mindset and tools to not only keep up with AI, but to lead within it. Insightful, practical, and deeply relevant."*

—DANIEL REEVES, Innovation Strategist, Keynote Speaker

"What sets The AI Shift *apart is its human-centered approach to artificial intelligence. This book emphasizes how individuals can use AI to improve their work, creativity, and quality of life. With contributions from 34 experts, it offers a rare combination of depth and accessibility. A must-read for professionals, leaders, and lifelong learners."*

—DR. HANNAH WELDRON, Adjunct Professor of AI, GRCC

"This book aims to help readers whether they've already 'answered the call' of AI or are just waking up to it. The AI Shift *is both an introduction and a roadmap, helping people understand not just what AI is, but how to use it to create meaningful results in their lives. It's an empowering guide for anyone who wants to thrive in civilization's next big disruption."*

—CASEY LINDBERG, Space X Quality Control Manager

THE AI SHIFT

THE AI SHIFT

Thriving in a World with Smart Machines and Robots

Authored by:
Erik Seversen, Daniel Abraão Carrasco, Lusine Aghajanyan, Samuel Akinyemi, Brie Bustos, Virbahu Nandishwar Jain, Kyle Cousin, Ervin Daniels, Dr. Cecilia Dones, DBA, Steffen J. Ehl, Roland Flouquet-Vilboux, Davide Gallo, Bart Gasiorowski, Leon Gelderblom, Vlad Grankovsky, Daniele Grassi, Prince Jain, Todd S. James, Alejandro Laplana, Brandon W. Ledford, Ron Ozminkowski, PhD, Sridhar Peddisetty, Stacy Proctor, Jenny Lopez Riley, Paul Robertson, DPhil, Dennis Sanchez, PsyD, Michael Schmid, PhD, Ryan Shatzkamer, Thomas Somogyi, Sakina Syed, Alex Szomora, Hatice Tavli, John Wei, Sam Wong, Haonan Wu

THIN LEAF PRESS | LOS ANGELES

Library of Congress Cataloging-in-Publication Data
Names: Seversen, Erik, Author, et al.
Title: *The AI Shift: Thriving in a World with Smart Machines and Robots.*
LCCN: 2026908370

ISBN 978-1-968318-59-8 (hardcover) | 978-1-968318-58-1 (paperback)
ISBN 978-1-968318-57-4 (eBook) | 978-1-968318-60-4 (audiobook)

Artificial Intelligence, Science & Technology, Business, Professional Development
Cover Design: 100 Covers
Interior Design: Dindo Sanguenza
Editor: Nancy Pile
Thin Leaf Press
Los Angeles

THIN
LEAF

Thank you for reading this book. There is information found within the following pages that can greatly benefit your life, but don't stop there. Make sure you get the most you can from this book and reach out directly to the expert-authors who want to help you navigate the AI shift so you can thrive in a world with smart machines and robots, and to manifest success in your life. Contact information for each author is found at the end of their respective chapter.

CONTENTS

INTRODUCTION xix
By Erik Seversen
Author of *Ordinary to Extraordinary*
Los Angeles, California

CHAPTER 1
MODERNIZATION IMPERATIVE: REINVENTING
THE ENTERPRISE FOR AN AI-DRIVEN FUTURE 1
By Daniel Abraão Carrasco
AI and Digital Transformation Thought Leader
São Paulo, Brazil

CHAPTER 2
AUGMENTED DECISIONS: KEEPING HUMANITY
AT THE HELM OF AI 11
By Lusine Aghajanyan
AI Human-Centric Strategy Leader
Nicosia, Cyprus

CHAPTER 3
AI AS A BRIDGE, NOT A BARRIER 21
By Samuel Akinyemi
AI Engineer and Entrepreneur
Lagos, Nigeria

CHAPTER 4
THE HUMAN THREAD 29
By: Brie Bustos
CIO, AI-Driven Strategist
Denver, Colorado

CHAPTER 5
THE INVISIBLE NINETY PERCENT 39
By Virbahu Nandishwar Jain
Founder & CEO, Quantisage
Hanover, New Hampshire

CHAPTER 6
THE 80/20 RULE OF AI TRANSFORMATION: WHY
ENTERPRISE AI IS 80% PEOPLE
AND 20% TECHNOLOGY..53
By Kyle Cousin
AI Executive
Boston, Massachusetts

CHAPTER 7
UNIFYING AI GOVERNANCE AND SECURITY: A
COMPREHENSIVE FRAMEWORK
FOR MANAGING AI RISK..61
By Ervin Daniels
Sr. Data & AI Platform Technical Specialist
Atlanta, Georgia

CHAPTER 8
THE HIDDEN SOCIAL SYSTEM
BEHIND EVERY AI SYSTEM...71
By Dr. Cecilia Dones, DBA
Professor, Founder, Chief Data Officer
New York, New York

CHAPTER 9
LEADING THROUGH THE SHIFT: HOW MODERN EXECUTIVES NAVIGATE AI-DRIVEN TRANSFORMATION 81
By Steffen J. Ehl
Business Executive and Advisory Board Member
Düsseldorf, Germany

CHAPTER 10
OPERATIONAL EXCELLENCE IN THE AGE OF AI 93
By Roland Flouquet-Vilboux
Tech-Enabled Industry Operating Leader & Advisor
Paris, France

CHAPTER 11
THE JOY OF AUGMENTATION: SCALING GO-TO-MARKET ENGINES IN THE NEW AGENTIC ENTERPRISE 105
By Davide Gallo
AI Leader, GTM Senior Director
Zurich, Switzerland

CHAPTER 12
THE SYSTEMATIC ADVANTAGE: MOVING AI FROM PILOT TO PRODUCT 115
By Bart Gasiorowski, MBA
Senior Executive, Strategic AI Programs & Entrepreneur
Orlando, Florida

CHAPTER 13
THE META-AGENT: OUR DIGITAL BUTTLER 127
By Leon Gelderblom
Sales Director Netherlands at HP Enterprises
Berkel en Rodenrijs, Netherlands

CHAPTER 14
THE 1% WHO WON'T LOSE JOBS 137
By Vlad Grankovsky
Machine Learning Expert and Entrepreneur
Hong Kong, China

CHAPTER 15
UNCOMFORTABLE QUESTIONS 145
By Daniele Grassi
CEO and Co-Founder at Axyon AI
Modena, Italy

CHAPTER 16
THE HUMAN EDGE 153
By Prince Jain
AI Technology Leader
San Jose, California

CHAPTER 17
NAVIGATING UNCERTAIN WATERS: LEADERSHIP LESSONS FOR THE AI ERA 163
By Todd S. James
CEO, Fortune 25 AI & Tech Executive
Cincinnati, Ohio

CHAPTER 18
OPERATIONAL AI AND THE RISE OF THE MID-MARKET ..173
By Alejandro Laplana
CEO at Shokworks
Austin, Texas

CHAPTER 19
HUMANITY AT THE CENTER OF AI & PUBLIC SERVICE 183
By Brandon W. Ledford
AI & Human Capital Transformation
Pittsburgh, Pennsylvania

CHAPTER 20
THE EVOLUTION OF HEALTHCARE AI..................................193
By Ron Ozminkowski, PhD
Founder, Analytic Strategies & Consulting, LLC
Ann Arbor, Michigan

CHAPTER 21
DECISION INTELLIGENCE: THE NEW OPERATING SYSTEM OF ENTERPRISE GROWTH 205
By Sridhar Peddisetty
Decision Intelligence & Agentic AI Leader
Wake Forest, North Carolina

CHAPTER 22
HUMANS & MACHINES TOGETHER: REDEFINING WORK IN THE AGE OF AI...217
By Stacy Proctor
Chief People Officer, AI Implementation Advisor
San Francisco, California

CHAPTER 23
THE NEW MARKETING MACHINE: DRIVING STRATEGY AND SPEED THROUGH TEAM TRANSFORMATION..229
By Jenny Lopez Riley
Chief Growth Officer, AI-Forward Leader
Chicago, Illinois

CHAPTER 24
A SELF-ADAPTIVE APPROACH TO BUILDING CREATIVE INTELLIGENT ROBOTS..241
By Paul Robertson, DPhil
AI Researcher
Boston, Massachusetts

CHAPTER 25
FROM THREAT TO THRIVE: INTRODUCING AI THRIVE HUMAN-AI PERFORMANCE FRAMEWORK 251
By Dennis Sanchez, PsyD
Founder of Perform/AI, Chief Human-AI Architect
Canton, Georgia

CHAPTER 26
THE LEADERS WHO WILL GET AI RIGHT (AND THOSE WHO WON'T) 259
By Michael Schmid, PhD
Founder of Fortify Intelligence; Scaling AI Safety
Boston, Massachusetts

CHAPTER 27
THE HUMAN LAYER: WHY AI SHOULD ASSIST, NOT REPLACE COMMUNICATION 275
By Ryan Shatzkamer
Senior Platform Engineer at outboundIQ
Boca Raton, Florida

CHAPTER 28
WHEN MACHINES DECIDE: TRUST AS THE INTERFACE OF THE AI SHIFT 287
By Thomas Somogyi
Head of Data, Automation, and AI in IT Consulting
Zurich, Switzerland

CHAPTER 29
THE RISE OF MICROENTREPRENEURS AND AIPOWERED SOLOPRENEURS 299
By Sakina Syed
Microsoft IT Professional, AI Engineer and Consultant
Toronto, Canada

CHAPTER 30
INVESTING IN INTELLIGENCE: WHY MONEY, WORK, AND GOVERNANCE WILL BE REWRITTEN 313
By Alex Szomora
Tech investor and Go-to-Market strategist
Singapore

CHAPTER 31
FROM COAL CAPITAL TO DATA CAPITAL: AI HALLUCINATIONS AND DIRE CONSEQUENCES.................325
By Hatice Tavli
Applied Mathematician & AI Inventor
Essen, Germany

CHAPTER 32
ENTERPRISE AI ADOPTION 333
By John Wei
CTO & SVP, Financial Services
Detroit, Michigan

CHAPTER 33
THE ARCHITECTURE OF VALUE: SOLVING BUSINESS OBJECTIVES THROUGH AI INNOVATION 343
By Sam Wong
Data & AI Builder, Innovator, Speaker and Instructor
Toronto, Canada

CHAPTER 34
ADAPT OR BE DELETED BY THE AI SHIFT: WHY HISTORY DOESN'T CARE ABOUT YOUR EXCUSES.............357
By Haonan Wu
Data & AI All-Rounder
London, England, United Kingdom

DID YOU ENJOY THIS BOOK?..369

INTRODUCTION

By Erik Seversen
Author of *Ordinary to Extraordinary*
Los Angeles, California

In a world of accelerating change, those who can reinvent themselves again and again will have the advantage.

—Yuval Noah Harari

A few days ago, when I was driving with my family, there was a car in the left lane of the freeway going slower than the flow of traffic. I put my turn signal on, shifted lanes, and passed the car. Also, for some reason recently, my dog wasn't finishing her food as consistently as normal. When I needed to get more dogfood, I purchased a different flavor. While I'm not sure how aware Elske was that I switched from lamb to salmon, the shift worked, and she began to finish her food again. On a slightly larger scale, when I entered college, I loved psychology and was planning on majoring in this field. However, after taking classes in literature, philosophy, and anthropology, I noticed that I was even more attracted to these disciplines, and I formally changed my major from psychology to anthropology. This shift had a large impact on the direction of my life, but it really was a minor shift in the big scheme of things.

Another shift example occurred just two days ago. I was on a short trip with my family as my kids were out of school for spring break. We were at a hotel with four of us sharing a room. At 4:50 a.m., I was jarred awake by the buzzing sound of my phone vibrating on a countertop across the room. I jumped up and went over to the phone. It was an incoming WhatsApp video call from a person I'd been wanting to speak with for a few weeks, yet he hadn't been getting back to me. As this individual is French, I wasn't sure if he'd want to speak English or French. I also imagined trying to work all of this out in a dark room, aiming to be as quiet as possible with my family sleeping in the room. My mind raced. It was an important call. Should I answer the video call in the bathroom, put on my clothes, and go down to the lobby and call him back? This early-morning call acted as a major shift within me as I had to transition from sleep to alert immediately and make decisions about an important moment.

Now imagine that everyone on the planet simultaneously got an important call, waking them up from sleep, and each had to make a quick decision on how to react. Some might take the call in the dark room with their pajamas on, some might find as good a spot as possible in the light of the bathroom to take the call, some might throw on some clothes, get to a more appropriate area, and call back right away, and others might decide to ignore the call, try to make contact at another time, and go back to bed, letting the results fall where they may.

When I think about how quickly AI is starting to affect many aspects of our lives (overtly and behind the scenes), I feel like this call in the middle of the night is what is happening to everyone in the modern world. Some have already gotten the call, and others are sleeping peacefully, unaware that the call is to arrive any moment, but the call will arrive.

This book, *The AI Shift*, isn't just about changing lanes on a freeway, buying a new brand of dogfood, or switching majors in college, it is about a transformation that is going to profoundly affect our lives. As the technology shifts, we need to wake up and create our own mindset shift as we begin to exist in a world where smart machines and robots become part of the landscape. One of the things I commonly say is "Ready or Not, AI is here." With this in mind, it is my goal to help as many people as possible be prepared for the shift that is taking place right in front of our eyes.

The intent of this book is not to scare people and focus on any doom and gloom regarding the future of AI, but rather to express that the AI shift is simply here and it is up to humans to decide how we are going to create the best possible future for ourselves and future generations.

Since I'm not an AI expert, I didn't try to write this book about AI by myself. Rather, I solicited the help of 34 AI experts from various backgrounds and locations. The co-authors of this book come from all over the USA, Canada, the United Kingdom, the Netherlands, Germany, France, Switzerland, Italy, Cyprus, Nigeria, Brazil, Singapore, and China. These authors have been featured on NBC, FOX, ABC, CBS, BBC, CSUN, and more. They are professionals who are university professors, scientists, business owners, consultants, product managers, engineers, software developers, cybersecurity leaders, healthcare experts, advisors, TEDx and keynote speakers, researchers, IT directors, CTOs, platform specialists, and more. The one thing these individuals have in common is that they all have something to share about artificial intelligence, and these ideas are available to you now.

Although this book is organized around the united theme of being prepared for the AI shift that is sweeping across our planet, each of the chapters is totally stand-alone. The chapters in the book can be read in any order. I encourage you to look through the table of contents and begin wherever you want. However, I urge you to read all the chapters because, as a whole, they provide a great array of perspectives. Each is valuable in helping you understand how AI is being used in many areas of human industry.

I hope that you discover something in this book that helps you navigate civilization's next big disruption as humans, smart machines, and robots become more interdependent and AI technology continues to insert itself into many aspects of our everyday lives. It is my hope that you embrace AI and you are able to create a mindset shift to keep up with, and be a leader in, the global transformations that are occurring.

Whether artificial intelligence becomes humanity's greatest advantage or its greatest challenge, AI is already reshaping the world around us. The real question is not whether AI will change our lives, but how we choose to engage with it. In this moment, we have a rare opportunity to partner with intelligent machines in ways that expand our capabilities, elevate our productivity, and create more space for the things that truly matter to us, like time with family. Never before have we held such powerful tools in our hands. Let's build a future where humans and AI work together to create a world that is filled with technological innovation yet still values creativity and human relationships. Like changing lanes or switching majors in college, let's shift with AI to put ourselves in the best position possible to reach our goals and live our best lives.

About the Author

Erik Seversen is on a mission to inspire people. He holds a master's degree in anthropology and is a certified practitioner of neuro-linguistic programming. Erik draws from his years of teaching at the university level and years of real-life experience in business to motivate people to take action, creating extreme success in business and in life.

Erik is a TEDx and keynote speaker who has reached over one million people through his public speaking and live courses. He has visited 99 countries and all 50 states in the USA and has climbed the highest mountains on four continents, 15 countries, and 18 states. Erik has published 20 bestselling books on the topics of mindset, success, and peak performance, and he has helped over 500 people become authors. He is a full-time writer, book consultant, and speaker, and he lives by the idea that success is available to everyone—that living an extraordinary life is a choice.

Erik lives in Los Angeles with his wife and has two boys currently studying at university.

Contact Erik for interviews, speaking, or book publishing consultation.

Email: Erik@ErikSeversen.com

Website: www.ErikSeversen.com

LinkedIn: https://www.linkedin.com/in/erikseversen/

CHAPTER 1

MODERNIZATION IMPERATIVE: REINVENTING THE ENTERPRISE FOR AN AI-DRIVEN FUTURE

By Daniel Abraão Carrasco
AI and Digital Transformation Thought Leader
São Paulo, Brazil

If everything seems under control, you're just not going fast enough.
—Mario Andretti, Formula One World Champion

Artificial intelligence will not replace humans, but it will replace every organization, professional, and executive who refuses to evolve.

This distinction matters. History shows that technological revolutions rarely eliminate people outright. What they eliminate are outdated ways of thinking, working, and creating value. The Industrial Revolution did not erase labor, it redefined it. The digital revolution did not end business, it reshaped it. The AI revolution follows the same pattern, but at unprecedented speed and scale.

The true disruption caused by AI is not technological. It is cultural, strategic, and deeply human. Algorithms are not dismantling careers. Stagnation is. Models are not replacing leaders. Obsolete leadership mindsets are. AI is accelerating a restructuring of civilization itself, forcing individuals and institutions to confront a fundamental question: How do we remain relevant in a world where intelligence is no longer scarce?

This chapter emphasises that the only sustainable path forward is reinvention. Not just of tools, but of how enterprises operate and how individuals learn, decide, and create value. AI is not a future event. It is the present infrastructure upon which the next era of human progress will be built.

AI as the New Infrastructure of Civilization

AI should not be understood as a tool. Tools are optional. Infrastructure is unavoidable.

Electricity transformed every industry without asking permission. The internet rewired communication, commerce, and culture simultaneously. Cloud computing dissolved physical boundaries between systems. AI now sits alongside these forces, not as an add-on, but as a foundational layer that reshapes everything it touches.

The most common mistake organizations make is treating AI as a project rather than a paradigm shift. They deploy chatbots, automate reports, or experiment with models while preserving pre-AI mental structures. The result is frustration. Not because the technology fails, but because the mindset does.

Organizations fail in the AI era not due to lack of access, funding, or data. They fail because they cling to outdated assumptions about work, decision-making, and authority. AI thrives in environments designed for learning, experimentation, and feedback. It suffocates inside rigid hierarchies and static processes.

For individuals, the shift begins with perception. AI is not a "new gadget" to test occasionally. It is a utility, like electricity. Once you see it that way, avoidance becomes irrational. Literacy becomes essential. This literacy is not about coding alone. It includes understanding data flows, automation logic, ethical implications, and how machine-generated insights interact with human judgment.

The most valuable professionals will be ambidextrous. They will combine human intuition, context, and values with machine intelligence. They will know when to trust data, when to challenge it, and when to override it. This balance, not technical mastery alone, defines relevance in the AI age.

From Automation to Augmentation: The Human Upgrade

When people ask me whether AI will take their jobs, I usually answer with another question: What part of your job actually makes you human?

AI is extremely good at automating tasks, that is not new. What is new is its ability to expand human thinking itself. The real shift we are living through is not automation, but augmentation. AI does not remove humans from the equation, it raises the bar for what humans are expected to contribute.

I have seen this pattern before. Tools never eliminate the need for intelligence; they redefine its expression. Calculators did not end mathematics. Spreadsheets did not destroy finance. They simply made superficial competence insufficient. AI is doing the same, only faster and across every profession at once.

What disappears are roles built entirely on repetition. What grows in value are capabilities rooted in judgment, synthesis, creativity, and emotional awareness. These are not "soft skills." They are survival skills.

Personally, I no longer see AI as something external. It has become part of how I think, write, analyze, and decide. Not because I outsource responsibility to it, but because it extends my cognitive reach. Used this way, AI feels less like a threat and more like a second brain.

The danger is not becoming obsolete because of AI. The danger is refusing to evolve while everyone else quietly upgrades.

Reinventing the Enterprise: From Hierarchies to Intelligent Systems

Most organizations I encounter are not failing because they lack AI tools. They are failing because they are structurally incompatible with intelligence that moves at machine speed. Traditional hierarchies were designed for predictability. The AI era is defined by volatility where decisions can no longer wait for layers of approval. Learning cannot be confined to quarterly reviews, data cannot live trapped inside silos.

Increasingly I see organizations as organisms rather than machines. Healthy ones sense their environment, adapt continuously, and learn from feedback. Unhealthy ones cling to static plans and rigid roles until reality overwhelms them.

Waiting for "the company" to modernize is, in my view, a mistake. Institutions move slowly. Individuals do not have that luxury. The professionals who thrive are those who modernize themselves first, who learn how AI connects to real outcomes, not just dashboards and demos.

Working today means collaborating not only with people, but with models, workflows, and automated systems. The ability to orchestrate this hybrid environment is becoming more valuable than any single technical skill. Relevance now comes from integration. From being the bridge between human intent and machine execution.

Winners and Losers: A New Digital Social Contract

AI is often accused of creating inequality, but I see it differently. AI exposes inequality that already existed. Access to tools is becoming cheaper. Access to understanding is not. The real gap is not technological, it is adaptive. Those who learn how to augment themselves accelerate, those who resist stagnate.

We are entering an era where the most meaningful divide will not be economic in the traditional sense. It will be between augmented and non-augmented humans, between those who extend their capabilities through AI and those who rely solely on manual effort. This is uncomfortable, but pretending otherwise is irresponsible. The response cannot be fear or denial. It must be intentional upskilling, continuous learning, and digital resilience.

Cybersecurity awareness, data privacy, and healthy reliance on intelligent tools are no longer optional. Neither is the shift from task execution to value creation. Tasks are abundant. Meaningful contribution is not.

The social contract is being rewritten in real time and participation requires engagement.

Leadership in the Age of Intelligent Machines

Much of what we still call leadership today was designed for a world that no longer exists. I see this gap clearly: executives operating intelligent systems with mental models shaped by the industrial age. The result is tension, confusion, and, often, quiet failure.

Experience has long been the currency of leadership. It still matters, but it is no longer enough. The pace of change introduced by AI has broken the assumption that past success can reliably guide future decisions. When environments evolve faster than experience can accumulate, leaders must supplement intuition with evidence, pattern recognition, and real-time insight. This is where AI becomes not a replacement for leadership, but its amplifier.

In this new context, leadership is less about control and more about orchestration. The role of the leader is not to compete with intelligent machines, but to decide how intelligence should be applied, where human judgment must prevail, and what values guide automated decisions. Authority no longer comes from knowing more than others, but from integrating diverse sources of intelligence into coherent direction.

I have learned that asking good questions is now one of the most strategic acts a leader can perform. Prompting AI is not a technical trick or a productivity hack. It is a reflection of how clearly one understands a problem. Leaders who cannot articulate their intent will be overwhelmed by outputs they do not trust or comprehend. Those who can frame questions well unlock clarity, alternatives, and foresight.

AI also changes how decisions are made. For the first time, leaders can explore multiple futures before committing to one. Scenarios can be simulated, risks surfaced, and assumptions tested in ways that were previously inaccessible. This does not eliminate uncertainty, but it replaces blind intuition with informed judgment. Decisions become less about betting and more about reasoning under uncertainty.

Trust, however, remains entirely human. In an AI-driven organization, credibility depends on transparency. People need to know where AI is used, how it influences outcomes, and where human responsibility begins and ends. Leaders who hide AI behind vague narratives erode confidence. Those who explain its role openly strengthen it.

Leadership in the age of intelligent machines is not about surrendering agency to algorithms. It is about using intelligence, human and artificial, in service of purpose. The leaders who succeed will be those who understand that technology does not define direction. Humans still do.

A Practical Roadmap for Survivors of AI Disruption

When people talk about surviving the age of AI, they often imagine something dramatic, as if adaptation were reserved for a small, technically gifted elite. My experience suggests the opposite. The ability to thrive in this environment is far less about brilliance and far more about intentional change.

What separates those who move forward from those who fall behind is not access to technology, but the decision to take personal modernization seriously. Organizations invest heavily in digital transformation, yet many individuals still treat learning as something occasional, triggered only by crisis. In an AI-driven world, that approach no longer works.

I have come to see learning as a system rather than an activity. Not an endless stream of information, but a deliberate cycle. Experiment regularly. Reflect on what works and what does not. Periodically redesign your skills and assumptions. This rhythm matters more than any single course or certification.

At the same time, how we present value is changing. Traditional resumes describe what someone has done. The AI era demands evidence of what someone can do now. Portfolios, projects, and visible demonstrations of AI-augmented thinking communicate relevance far more clearly than static credentials ever could.

This shift also requires a deeper mental reframing. The question "What is my job?" is becoming increasingly fragile. Roles change faster than identities can keep up. A more resilient question is "What value do I create, and for whom?" Value travels more easily than titles. It adapts when structures collapse.

AI disruption is not a temporary wave that will pass. It is a permanent condition of modern life. Accepting this is not an act of resignation, but of freedom. Once you stop waiting for stability to return, you gain the ability to move, learn, and reinvent on your own terms.

Survival, in the end, is not about resisting change. It is about building the capacity to change continuously.

Conclusion: Humanity Reinvented

AI has a way of forcing uncomfortable questions to the surface. Questions about work, identity, creativity, and even meaning. For many, this feels destabilizing, as if something essentially human were being challenged. I see it differently. I see a mirror being held up to us, asking what truly matters when intelligence itself is no longer scarce.

For most of history, human value was tightly coupled to our cognitive limits. We memorized, calculated, and processed information because machines could not. That constraint shaped how work was organized and how people defined their worth. AI breaks that constraint. And when constraints disappear, assumptions must be revisited.

This moment is unsettling precisely because it removes familiar anchors. If machines can think, analyze, and generate at scale, what remains uniquely human? The answer is not found in competing with algorithms, but in leaning into what they cannot replicate easily: imagination, ethical judgment, empathy, and the ability to create meaning from complexity.

Technology will not determine our future on its own. It never has. The direction we take depends on our willingness to adapt not just our tools, but ourselves. The greatest advantage in the decades ahead will not belong to those with the most advanced systems, but to those who are capable of continuous reinvention.

I believe the real invitation of the AI era is this: to let go of identities built around repetition and efficiency, and to reclaim the deeply human dimensions of work and life. Curiosity becomes more valuable than certainty. Learning becomes more important than mastery. Flexibility outweighs control.

My advice, then, is not to fear AI, nor to idolize it. Approach it with openness. Revisit your skills every few years with honesty. Allow machines to absorb what is mechanical and predictable, so you can invest more of yourself in what is creative, connective, and transformative.

The future shaped by AI is not artificial. It is a future that asks us, again and again, to become more human.

About the Author

Daniel Abraão Carrasco is a technology executive, strategist, and educator with more than three decades of experience operating at the intersection of digital transformation, cloud computing, and artificial intelligence. Based in São Paulo, Brazil, he has built a global career leading large-scale modernization initiatives across Latin America and working closely with multinational organizations during periods of deep technological change.

Daniel began his career in the early 1990s at IBM, where he developed a strong technical foundation in enterprise systems, infrastructure, and large-scale IT operations. Over the years, he held senior and executive leadership roles at global technology and consulting companies including IBM, HP, Dell, Unisys, Accenture, and Avanade. Throughout this journey, he consistently led complex organizations, global delivery teams, and strategic portfolios with full P&L responsibility.

Academically, Daniel brings a multidisciplinary foundation that bridges technology and business. He holds a degree in computer science from IMES University in São Paulo and a degree in business administration from Universidade Metodista de São Paulo. He later completed postgraduate studies in software engineering at the University of São Paulo (USP) and a specialization in agile methods at Fundação Getulio Vargas (FGV São Paulo).

Currently, Daniel serves as CTO and managing partner for digital business at Think IT, where he leads cloud operations and portfolio strategy for large-scale digital transformation programs. In parallel, he is the founder and CTO of technology ventures focused on enterprise platforms, AI-enabled business systems, and the acceleration of technology careers. His leadership style connects people, processes, and platforms to increase resilience, efficiency, and long-term value creation.

In parallel with his executive career, Daniel is actively engaged in academic teaching. He serves as a professor of multicloud architecture at FIAP, one of Brazil's most recognized institutions for technology and digital innovation, where he teaches practical and strategic applications of AWS, Azure, and Google Cloud Platform to the next generation of technology professionals.

Fluent in English, Portuguese, and Spanish, Daniel is a frequent speaker and published author on digital transformation, artificial intelligence, and the future of work.

This chapter reflects both his academic grounding and his lived experience leading technology transformations from inside organizations. It is written for leaders and professionals who understand that the AI era is not primarily a technological challenge, but a human one.

Email: dqcarrasco@gmai.com

LinkedIn: https://www.linkedin.com/in/daniel-carrasco-cloud/

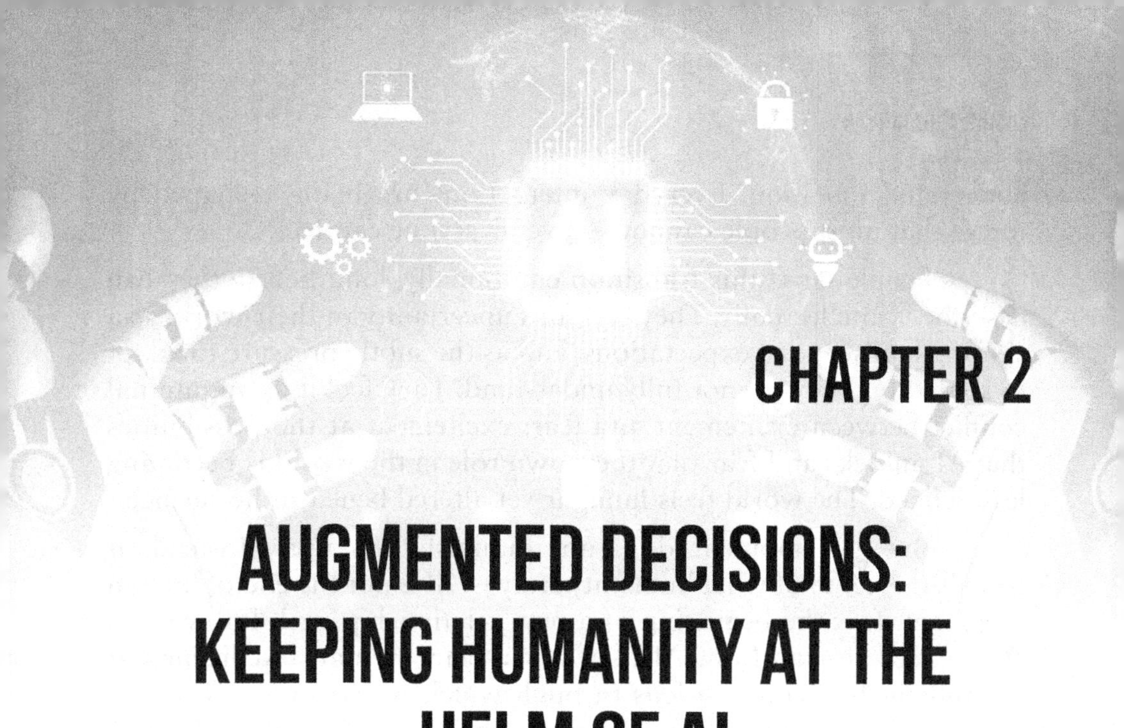

CHAPTER 2

AUGMENTED DECISIONS: KEEPING HUMANITY AT THE HELM OF AI

By Lusine Aghajanyan
AI Human-Centric Strategy Leader
Nicosia, Cyprus

Between stimulus and response, there is a space. In that space is our power to choose our response.

—Viktor E. Frankl

Humanity stands inside that space between stimulus and response today. Artificial intelligence has become the stimulus that is reshaping the architecture of our daily lives, often quietly and without dramatic announcement. The world has not changed through a sudden revolution, but through a slow expansion of capability. Systems that once delivered simple answers now analyze, interpret, and anticipate. Tools that once waited for human instruction now respond as if they understand intention. Everyday interactions are being reshaped by forces that most people cannot see, yet everyone can feel.

People sense this transition emotionally, long before they can describe it intellectually. They feel it as uncertainty in their work, as a shifting landscape of expectations, and as the subtle pressure to adapt to something they cannot fully understand. They feel it as an internal conflict between excitement and fear: excitement at the possibilities that AI unlocks and fear that their own role in the world is becoming less defined. The world feels familiar yet altered beneath the surface.

But this moment does not diminish humanity. Instead, it reveals the depth of what humanity truly is. AI is not the end of human judgment. It is the beginning of a new relationship with knowledge. When used thoughtfully, AI does not replace human intelligence. It expands it. It becomes a lens through which we see more clearly, a light that illuminates what we could not see alone. AI widens the field of possibility. Humanity chooses the purpose.

This is the heart of augmented decisions. They are born from a partnership between machine clarity and human meaning. Intelligence evolves, but the responsibility for direction remains human. A system can analyze data, but only a person can interpret what that analysis means for a life, a community, an organization, or a future.

To understand this partnership, we must remove the illusion surrounding AI. Despite its sophistication, AI remains straightforward in essence. It is a system that learns patterns. Every digital record—every sentence typed, every image captured, every movement tracked—becomes part of its understanding. AI consumes these fragments of experience and discovers repetition. From repetition, it predicts. It sees possibility but not purpose. It sees pattern but not people.

Humans, however, understand meaning. Humans know why a choice matters. Humans can feel the emotional impact that does not live inside data. Humans understand nuance, compassion, intuition, and responsibility. These are not technical details. They are the anchors of humanity.

This is why AI must remain a tool rather than a guide. Its intelligence expands perception but cannot define meaning. It reveals what might happen but not what should happen. Without humans, the intelligence of machines becomes directionless. Without machines, the intelligence of humans becomes limited. Together, they create a deeper form of decision-making, one that blends insight with conscience, clarity with care.

To see this clearly, consider how decisions form. Most people view decisions as singular events, but they are chains of influence. They begin with what we choose to observe. They evolve as we interpret those observations. They absorb emotion, history, culture, and experience. They take shape through values. When AI joins this chain, it becomes one link among many, not the beginning, not the end, but a contributor. It can reveal patterns and possibilities we could not see alone. But humans choose meaning. Humans choose direction. Humans choose consequence.

Without human involvement, decisions lose their grounding. A model trained on biased data repeats injustice. A metric chosen without intention optimizes outcomes that conflict with values. A prediction presented without explanation becomes mistaken for truth. A decision made without reflection becomes mechanical. Augmented decisions prevent this by anchoring the entire chain in human conscience.

When done thoughtfully, AI enhances rather than replaces. In healthcare, AI can detect anomalies long before symptoms appear, but healing remains a human dialogue, a space where fear meets empathy, where uncertainty meets wisdom. In education, AI may identify gaps in knowledge, but a teacher understands the emotion behind a child's silence or the insecurity hidden beneath a sudden loss of confidence. In business, AI can optimize, forecast, and accelerate, but leaders determine what is ethical, fair, sustainable, and aligned with human dignity.

As AI grows more capable, humans face an internal evolution that few have prepared for. The shifts happening in society are not only technological; they are psychological. For centuries, society defined intelligence by the functions of the human mind: reasoning, memory, calculation, decision-making. Now machines participate in these functions, and people naturally wonder what remains exclusively human. They feel a subtle fear: If machines can think in certain ways, what is left for us?

This is not fear of AI. This is fear of losing identity.

People do not fear machines replacing their work. They fear machines replacing their sense of worth. They fear becoming invisible inside systems that cannot feel them. They fear being asked to keep pace with a world that evolves too quickly for their own emotional rhythm. These fears are not weaknesses. They are signaling that humanity is being asked to evolve.

And evolution begins with awareness. AI pushes us toward deeper parts of ourselves, parts that were not always required in the same way. It pushes us to develop emotional intelligence, reflection, intuition, and resilience. These traits do not compete with AI. They complete it. AI expands external intelligence. Humanity must expand internal intelligence.

The leaders who will guide this era are those who understand both intelligence and emotion. They will not be chosen for how much they know, but for how deeply they perceive. They will be the ones who bring calm into complexity, clarity into uncertainty, and compassion into systems that do not naturally understand human experience. These leaders are the emerging architects of human-centered decision-making.

Their role is not limited to technology. Their role is to protect meaning, something AI cannot measure or predict. Meaning is created by humans. It emerges from relationships, purpose, responsibility, and dignity. Without meaning, intelligence becomes directionless. Without purpose, systems become unstable.

My own understanding of this came long before AI became central to my work. For over fifteen years, I worked in global maritime leadership, a world defined by uncertainty, responsibility, and the constant presence of human lives relying on decisions. Maritime leadership teaches calm in chaos. It teaches balance between technical knowledge and human intuition. It teaches how to read systems and people simultaneously. It teaches that the greatest decisions arise not from certainty but from courage. These lessons became the foundation of how I understand AI, not as a competitor but as a companion to human judgment.

When I entered the world of intelligent systems, I saw the same patterns of responsibility but magnified. Technology was advancing faster than people's understanding. Leaders felt overwhelmed. Employees felt threatened. Teams felt unprepared. People were not afraid of machines. They were afraid of decisions being made without their voice, without their context, without their humanity.

My mission emerged naturally: to stand between humans and intelligent systems as a translator, a guide, a protector of human-centered decision intelligence. To help people understand technology not as a threat but as an opportunity for expansion. To help leaders use AI without losing the emotional foundations that make leadership meaningful. To help organizations grow without sacrificing human dignity. To ensure that the evolution of technology is matched by an evolution of humanity.

As AI evolves, it does not ask us to think faster. It asks us to think deeper. It asks us to become more aware of the emotional and ethical structures behind our decisions. It asks us to expand the capacities that machines cannot touch. It asks us to evolve into more conscious versions of ourselves.

The world does not need humans who imitate computers. It needs humans who elevate consciousness. It needs humans who remain grounded in purpose. It needs humans who understand that wisdom is not speed. Wisdom is presence.

The presence of AI forces humanity to confront a unique mirror. It reflects to us not only our intelligence, but our assumptions, our blind spots, and our unresolved relationship with responsibility. Every limitation AI reveals is an invitation for human growth. Every bias discovered in a model points toward a societal truth that must be acknowledged. Every misinterpretation exposes the places where emotional context is missing. AI does not erase humanity. It clarifies where humanity must evolve.

This moment in history asks something profound of us. It asks us to step beyond the narrow definitions of intelligence we once held. For generations, intelligence was measured by memory, logic, and speed. Today, machines surpass us in all three. But these were never the full measure of human intelligence. Humanity's greatest strengths have always been our ability to reflect, to feel, to interpret nuance, and to act with purpose even when information is incomplete. These strengths are not diminished by AI. They are amplified by it.

AI compels us to awaken the inner capacities that define who we are: the capacity to sense meaning beneath information, the capacity to feel responsibility beyond logic, the capacity to hold empathy in the presence of uncertainty, the capacity to imagine futures that data alone cannot predict. These capacities do not make us slower. They make us deeper.

As systems grow more intelligent, humanity must grow more conscious. Consciousness is not a technical term. It is the ability to understand reality not only through information but through insight. Consciousness transforms knowledge into wisdom. It brings awareness of consequences beyond what is measurable. It brings ethical grounding into systems that have none. It brings emotional maturity into decisions that affect more than performance indicators. Consciousness is humanity's greatest contribution to the future of intelligent systems.

This is why decision-intelligence leadership is no longer optional. We need leaders who can interpret patterns while understanding people. Leaders who can explain complexity without overwhelming others. Leaders who can make decisions that honor both data and dignity. Leaders who remain calm when the world accelerates and compassionate when people feel left behind. Leaders who support teams not only in using new tools, but in understanding how those tools think.

The world does not lack technology. It lacks people who know how to humanize technology. It lacks people who see AI not as a competitor but as a catalyst for human evolution. It lacks individuals who can guide organizations through transformation without losing sight of the people inside them.

Every era of progress has required a new type of leader. The industrial era required builders. The digital era required innovators. The AI era requires integrators, individuals who bridge human depth and technological intelligence. This bridge is not built by expertise alone. It is built by presence, awareness, responsibility, and the courage to defend human needs in environments driven by data.

My maritime years taught me the foundations of this leadership long before I ever named it. At sea, responsibility is immediate. The environment shifts without warning. Decisions carry human consequence. A leader must remain calm under pressure, intuitive in uncertainty, and deeply attuned to the people who rely on them. This balance between technical clarity and human depth shaped me. It taught me that systems alone are never enough. People are always at the center, even when the system is large and complex. AI has the same structure. It can guide, but it cannot feel. It can inform, but it cannot care. It can illuminate, but it cannot choose.

When I entered the world of AI, I saw the same patterns magnified on a global scale. Organizations were adopting new systems, but people were not being prepared emotionally. Teams were expected to adapt without understanding. Leaders were being asked to make decisions influenced by models they did not fully trust. The gap between capability and comprehension widened. This gap is not technical; it is human. And it became clear to me that my mission was to help close it.

Helping people work with AI is not simply about explaining how tools function. It is about reducing fear. It is about restoring clarity. It is about reminding people that they still matter, perhaps more than ever. It is about guiding them to see that their humanity is not competing with technology. It is shaping it. The future needs humans who understand themselves deeply enough to guide systems designed to think quickly, but not meaningfully.

As AI evolves, humanity is being invited to evolve with it. We must develop emotional resilience to face ambiguity. We must develop inner grounding, so we are not shaken by rapid change. We must strengthen our intuition, because not every decision can be reduced to logic. We must deepen our ethical understanding, because systems have no conscience. And we must expand our capacity for empathy, because the more automated the world becomes, the more humans will rely on emotional connection.

This evolution is not theoretical. It is already happening in every industry, every company, every team. It is happening in quiet conversations between leaders and employees. It is happening in boardrooms where questions arise about what deserves automation and what must remain human. It is happening in families where children interact with intelligent tools long before they understand their implications. It is happening in society as people seek clarity about their future.

AI does not reduce humanity. It calls humanity forward.

The future will belong to those who can merge intelligence with consciousness, capability with compassion, and progress with purpose. It will belong to leaders who understand that clarity is not only intellectual but emotional. It will belong to those who recognize that transformation is not only technological but human. It will belong to those who bring depth into environments driven by speed.

And so, we return to the metaphor that has guided much of my thinking. Imagine yourself standing on the deck of a ship at night. The sea stretches endlessly around you, dark and unknowable. You switch on a powerful beam of light. Suddenly the darkness becomes detail. Hidden rocks appear. Safe passages form. The horizon gains shape.

But the beam does not steer the ship. It does not feel the responsibility in your chest. It does not sense the trust of the people on board. It does not weigh the meaning of the journey, nor the future it leads to.

The light reveals the world. But only you can choose the path through it. AI is the light. Humanity is the helm. The light can reveal the way. Only the human can choose it. This is augmented decision-making. This is the next era of leadership. This is the bridge between technology and meaning. This is the evolution of humanity in an intelligent age.

The future will not be shaped by how smart machines become, but by how conscious humans choose to be. Intelligence will continue to expand. It is up to us to ensure that meaning expands with it. As systems evolve, we must evolve our wisdom. As tools grow powerful, we must grow purposeful. And as AI becomes a partner in our decisions, we must remain the authors of our direction.

The helm is still ours. We move forward with clarity. We move forward with courage. And we move forward with humanity at the center of every decision.

About the Author

Lusine Aghajanyan is a strategist and decision-intelligence engineer dedicated to elevating human-centered leadership in a world increasingly shaped by intelligent systems. With more than 15 years in global maritime operations, she learned how high-uncertainty environments demand both analytical precision and profound emotional depth. These experiences shaped her belief that meaningful transformation requires leaders who understand people as deeply as they understand data.

She is the founder of Lunova: decision-intelligence practice guiding organizations through data-driven evolution without losing their human core. Her work focuses on removing fear from AI, helping individuals understand not only how intelligent systems operate but how they can be used to protect dignity, strengthen responsibility, and amplify long-term purpose. Her mission is to cultivate a new generation of leaders who can translate human needs into technology and ensure that, as systems advance, humanity remains at the helm.

Email: La@Lunova.global

Website: https://lunova.global/

LinkedIn: https://www.linkedin.com/in/lusine-aghajanyan-la13/

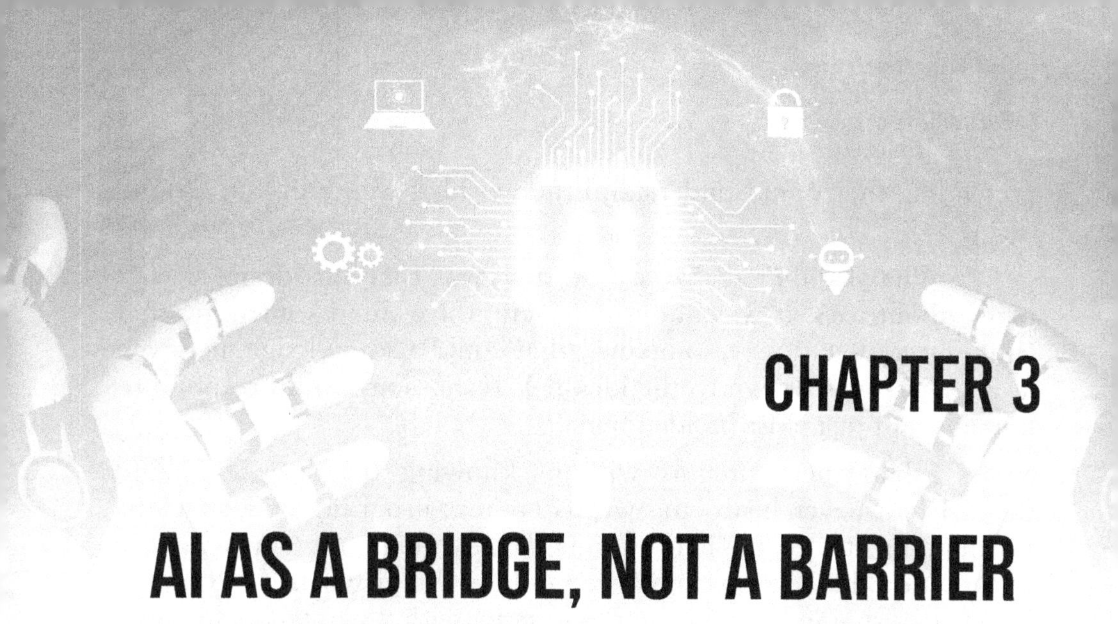

CHAPTER 3

AI AS A BRIDGE, NOT A BARRIER

By Samuel Akinyemi
AI Engineer and Entrepreneur
Lagos, Nigeria

Technology is best when it brings people together.
—Matt Mullenweg

Long before I had any knowledge of coding or could even explain what artificial intelligence meant, I was already imagining a world filled with robots. As a child, I dreamed of creating machines that could think, play, and interact with kids, almost like a massive toy garden where imagination and technology met. I didn't know how such things could be built, but I felt strongly that they were possible.

Back then, all the toys I owned were silent. They never acted on their own, never spoke back, and never responded in any meaningful way. When we played with them, we had to supply the voices, the movement, and the stories ourselves. Yet even in that simplicity, I wanted more. I wanted toys that could respond, that could understand, and that could feel alive in some small way. I remember lining them up and pretending they were listening, pretending they could react, pretending they could understand the worlds I was building around them.

What I didn't realize at the time was that this desire wasn't really about toys. It was about connection. I wanted something that could meet me halfway, something that could acknowledge me. Even as a child, I was drawn to the idea that technology could respond to humans, not just exist beside them.

I had these dreams without knowing the science behind them. I had never heard the words coding, programming, machine learning, or algorithm. Those terms did not shape my curiosity. My imagination did. I wasn't interested in how things worked internally; I was interested in how they felt to interact with. I wanted technology that didn't feel cold or distant, but present.

When I began my formal journey into computer science, those terms remained distant and unfamiliar to me. My dreams were never shaped by textbooks or technical definitions, but by imagination and curiosity. I did not understand how machines worked, only this: I wanted to create something that could think, respond, and interact with people in a meaningful way.

Looking back now, I realize that those early dreams were not really about machines at all. They were about possibility. They were about creating tools that could reduce loneliness, spark curiosity, and make people feel less alone in the world. In many ways, they were early sketches of a philosophy I would later come to embrace: Technology should bring people closer, not push them apart.

Let's Take a Walk Back in Time

On November 30, 2022, Sam Altman posted a simple message on X (formerly Twitter) announcing that ChatGPT was live. That single tweet marked a turning point. For the first time, millions of people could speak directly to artificial intelligence. In that moment, AI shifted from something technical and distant into something personal and accessible. It was no longer hidden inside research labs or complex systems. It was now in front of everyday people, ready to listen, respond, and collaborate.

What made that moment powerful wasn't just the technology itself, but the feeling it created. People weren't reading about AI anymore; they were conversing with it. Students asked questions. Writers explored ideas. Professionals experimented. For many, it was the first time technology felt less like a tool and more like a participant.

However, that tweet did not create artificial intelligence. It revealed it. What felt sudden to the world was actually the result of decades of research, experimentation, and imagination finally becoming visible. What appeared like an overnight breakthrough was in truth the product of a very long journey.

As I deepened my understanding of artificial intelligence through an introductory course at Harvard University, I realized that AI had always been part of the story of computing itself. Long before artificial intelligence had a name, pioneers like Alan Turing were already asking the most important question back in the 1950s: Can machines think? That question reshaped the direction of technology. It transformed computers from simple calculating tools into systems designed to imitate logic, reasoning, and human thought.

In many ways, every computer built since then has carried an ambition to move closer to intelligence. Each generation of hardware and software brought machines a little closer to understanding patterns, language, and decisions. What we witnessed in 2022 was not the birth of AI, but the moment it crossed a bridge into public consciousness.

If OpenAI gave the world a voice to speak with artificial intelligence, NVIDIA gave AI the power to think at scale. Long before AI became conversational, NVIDIA had already built the computational backbone that made large language models possible. By transforming graphics processors into engines for deep learning, they enabled machines to process massive amounts of data at extraordinary speed. Their work quietly powered the AI revolution long before it had a public face.

Alongside them, many others shaped this era. Research labs, universities, and open-source communities across the world laid the foundation. What we see today is not the work of one company, but a shared human effort that turned artificial intelligence into a global reality.

Startups as Living Proof of AI as a Bridge

In the years since AI burst into the public eye, one of the most inspiring places I've seen its bridging power is in the world of startups. In co-working spaces from Lagos to Bangalore, I've met young founders eager to show how their AI-powered ideas can bring people closer together. Their eyes light up describing how a simple app can connect a farmer to fair market prices or pair a student in a remote village with a world-class tutor.

These entrepreneurs aren't household names, but their work is quietly removing distance, friction, and inefficiency in people's lives. They aren't building for headlines. They're building for people who were previously excluded from opportunity.

Take, for example, a small education platform I encountered during a community tech fair. The team had built an AI tutor that works offline, guiding students through lessons in their local language. I watched a shy teenager solve math problems with the tutor's gentle prompts and encouraging feedback. This wasn't just a clever gadget. It was a bridge. The AI removed the barrier of limited teachers at her rural school, letting her learn at her own pace. It felt personal and empathetic, as if the software understood what she needed.

I've seen similar bridges being built for creatives and freelancers. As a software engineer in a small suburb of Lagos, I gained access to international clients through Upwork's matching algorithm. The AI quietly matched me with projects that fit my skills, removing the exhaustion of endless searching. It felt like a door opening, one I had always known was there, but never had the key to.

These stories reveal something important: When technology is designed with empathy, it doesn't dominate human lives. It supports them. AI becomes less about automation and more about empowerment. Less about replacing effort and more about amplifying potential.

When AI Becomes a Barrier

Yet not every AI system lives up to this promise. I've also seen technology that, instead of connecting people, ends up isolating or confusing them. Often, the culprit is poor design or misplaced intention.

I once spent an hour arguing with an automated customer service bot that couldn't grasp my problem. It repeated scripted responses while ignoring context. In that moment, the AI wasn't a helpful bridge. It was a wall. It blocked access to human understanding instead of facilitating it.

There are times when AI tries to replace human thinking instead of supporting it, and the result can be troubling. I remember speaking with an educator who noticed her students relying on AI tools to write their essays. The output looked polished, but the thinking behind it was missing. The students weren't learning to reason, question, or struggle with ideas. The AI became a shortcut, and in doing so, quietly eroded growth.

Worst of all, automation can sometimes strip away dignity. Workers replaced without retraining. Decisions made by algorithms that no one can question. Systems that unintentionally reinforce bias. These moments remind us that AI is not neutral. It carries the values, assumptions, and blind spots of those who build it.

Recognizing these risks isn't about rejecting AI. It's about demanding better. It's about ensuring that innovation does not come at the cost of humanity.

Designing AI for Human Connection

So how do we make sure AI remains a bridge and not a barrier? Over time, I've come to believe it requires intentional design rooted in empathy. The most impactful AI systems I've encountered share a few core principles: simplicity, accessibility, transparency, and inclusion:

- Simplicity means technology should feel intuitive. The best AI tools disappear into the background. They don't demand attention; they enable focus.
- Accessibility means AI must work for people with slow internet, older devices, or personal disabilities. If a tool only serves the privileged, it fails its purpose as a bridge.
- Transparency builds trust. People deserve to know when AI is involved and why it behaves the way it does. Trust grows when systems are understandable.
- Inclusion ensures that technology respects language, culture, and identity. AI should not force people to conform. It should adapt to them.

I've always been struck by how science fiction imagined this long before we had the tools to build it. In TV shows like *Star Trek* or films such as *Space Sweepers*, people speak freely in their own languages, yet everyone understands each other in real time. The technology never asks them to change who they are. It stays in the background, quietly adapting, acting as a bridge rather than a spotlight.

Ultimately, designing AI for human connection means putting empathy ahead of efficiency. Just because something can be automated doesn't mean it should be. The real question is not "What can AI do?" but "How does it make people feel?"

Bridging Communities in the Real World

When I think about AI's potential, I often picture the communities left behind by previous technology waves. In these places, a single smartphone paired with thoughtful AI can change lives.

I've seen nurses in rural clinics use AI tools to assist diagnoses where doctors were unavailable. I've seen farmers consult AI systems in their own languages to protect crops. These aren't futuristic fantasies. They are quiet revolutions.

Accessibility here isn't optional. It's essential. If AI demands high-speed internet or high costs, it excludes the very people who need it most. But when built intentionally, AI becomes a bridge to education, healthcare, and opportunity.

This places responsibility on founders and builders. Every system we deploy shapes culture. Every design choice sends a message about what and who matters. The founders I admire ask deeper questions before launching. They understand that trust is harder to build than software and more valuable in the long run. In the end, technology exists to enhance human life. Before AI becomes powerful, it must be human.

About the Author

Samuel Akinyemi is an AI and full-stack engineer, entrepreneur, and founder of Sannex Tech Ltd, where he leads the development of multiple technology startups across ed-tech, AI, fintech, automation, and digital infrastructure.

Samuel works at the intersection of machine learning, large and small language models, and scalable software systems, building products that solve real business and societal problems. Actively engaged in startup formation and early-stage product development, Samuel has helped launch and scale platforms used by students, freelancers, and small businesses across emerging markets.

Samuel is passionate about making AI practical and accessible, believing smart machines should expand human capability rather than replace it. Beyond building companies, he mentors young developers and founders, helping them navigate technology, entrepreneurship, and responsible innovation.

Email: samuel@sannex.ng

Website: www.sannex.ng

LinkedIn: https://linkedin.com/in/sam-akins

CHAPTER 4

THE HUMAN THREAD

By: Brie Bustos
CIO, AI-Driven Strategist
Denver, Colorado

We expect more from technology and less from each other.
—Sherry Turkle, *Alone Together*

The AI shift is often framed as a technological revolution. It is, but at its core, it is a human one. From humans being the ones to create the technology, to being the ones to test, implement, and support the technology that will ultimately drive enhancements now and into the future. On the other side of that is people all too ready to give up our own agency, to outsource the most fundamental parts of being a human: creativity, connection, learning, tenacity. The rapid expansion of AI has not been accidental. It has been fueled by our appetite for convenience and amplified by markets that reward anything capable of scaling attention, efficiency, or profit. This is not meant to be a cautionary tale, but perhaps it should be.

This isn't just algorithms, frameworks, and infrastructure driven by ones and zeros. It is driven by heartbeats. The heartbeats of people in their homes, asking questions, seeking reassurance, and shaping daily habits. The heartbeats of leaders guiding organizations through technological change unlike anything we have seen before. Once imagined in futuristic films, we are now living through an era that is reshaping not only how we work, but how we relate, decide, and connect.

What happens when we overlook the human layer that threads through this explosive technology and people?

Let's start with the technology. Humans built it and trained it with human knowledge, and we continue to use it at an exponential rate. The goal would be that the technology being built must be useful to humans. If humans aren't driving this technology with intention, there is no future for AI that truly benefits us.

The primary consumer technologies shaping how people interact with AI today are large language models, chat interfaces, and AI-enhanced search tools. According to a March 2025 national survey conducted by Elon University's Imagining the Digital Future Center, approximately 52% of US adults report using AI large language models such as ChatGPT, Claude, and Gemini with broader estimates placing total AI tool usage between 50% to 60% of American adults. These tools are not used solely for information retrieval; many people engage them for conversational and relational purposes. These are interactions that feel increasingly human in tone and structure. These tools are functioning as their therapist, their friend, their personal trainer and coach, even their spouse. We are replacing humans with AI that is attempting to replace human connection above anything else.

Beyond language models and chat interfaces, AI is embedded in everyday life through smart home technologies. According to recent adoption data, nearly half of American households now have at least one smart home device, including smart speakers, connected lighting, thermostats, and security systems. These systems operate quietly in the background, reminding us to shut the garage door, adjusting temperature automatically, playing music on command, or managing daily routines.

Consumers consistently adopt these devices for convenience, comfort, and efficiency, but the cumulative impact extends beyond mere task automation. Smart thermostats learn household patterns; voice assistants respond conversationally; sensors adjust lighting and temperature based on behavior. Over time these systems do more than execute commands. They anticipate preferences and shape expectations about responsiveness and personalization.

When a device adjusts lighting to match mood, queues music based on past preferences, or answers questions with a reassuring tone, it begins shaping more than efficiency. It begins influencing ambiance and, by extension, emotional experience. We grow accustomed to immediate responsiveness and frictionless interaction. In subtle ways, we use these systems not only to manage tasks but increasingly to co-regulate our environment and, at times, even aspects of our mood and interpersonal space.

That's why it can feel like the most responsive "presence" in the room is the device itself, not because the technology has replaced human depth, but because it is always available, always attentive, and never fatigued, occupying relational space that once depended on human attentiveness and presence.

We have offloaded very human connections to computers when maybe we would be better off if the technology took over the mundane tasks in our lives. I, for one, would gladly sign up for a robot that cleaned my house, did my dishes, including putting them away, did my laundry from start to finish, including, let's be honest, putting it away, put gas in my car, and changed my oil, just to name a few. What would I give to spend a weekend on the mountain slope or on the lake with friends and family knowing I was coming home to a clean and tidy house? That is technology I would sign up for.

Even in business, the reality of AI today is pragmatic and unglamorous. Research consistently shows that the most widely adopted enterprise AI applications are focused on routine, high-volume, human-facing tasks rather than deep strategic augmentation of complex work.

According to multiple 2025 industry surveys, the top business AI use cases include:

- Customer service automation: AI chatbots and virtual assistants handling common inquiries, improving response times, and deflecting routine tickets
- Intelligent email and ticket triage: systems that categorize, route, and sometimes respond without human intervention
- Agent assist tools: AI supporting frontline agents by generating responses or summarizing context in real time.

These tools are effective at handling high volume, predictable interactions, which means they touch some of the most human-centric parts of business communication.

But look at what is not dominating AI investment:

- Tools specifically designed to synthesize complex executive strategy across stakeholders
- Automated systems that resolve systemic operational friction such as infrastructure bottlenecks (e.g., network latency or platform performance)
- AI systems built to analyze internal spend and drive human-centered cost savings that could be redistributed into workforce well-being or profit sharing.

In practice, AI adoption has not started with transforming the hardest parts of corporate life; it's started with the most visible, measurable, and monetizable parts. And that's no accident.

The rapid expansion of enterprise AI tools did not occur in a vacuum. It occurred in response to our demand for speed, availability, personalization, and reduced friction. Research on enterprise technology adoption consistently finds that:

- Efficiency gains and cost reduction are the top drivers of AI deployment, often outpacing goals like job augmentation or human enablement.

- Reducing customer service workload is cited as a leading business priority, ahead of internal innovation or workforce development.

In a capital-driven market, incentives align toward what can be scaled, quantified, and monetized. Governments and boards aren't funding AI pilots to make human lives richer. They're funding AI pilots to make balance sheets look stronger. That dynamic inevitably leads to the question many executive teams quietly ask: If this technology can perform that function, how many people can we do without?

And because workforce cost is one of the largest controllable levers in operating budgets, replacing headcount becomes not just a possibility but a measurable outcome under evaluation.

The machine does not originate meaning; it scales what it is given. The quality of AI output is directly tied to the quality of human input. That is the human layer, the thread that binds technological advancement to human capability. As AI systems grow more powerful, the need for human depth does not diminish; it intensifies.

The human layer is not abstract. It shows up in how leaders make decisions when the data is incomplete. It shows up in how managers communicate change. It shows up in how teams collaborate across disagreement, interpret nuance, and navigate ambiguity. It exists in judgment calls that cannot be fully modeled, in ethical considerations that do not appear in dashboards, and in the lived context that no dataset fully captures.

In organizations, the human layer determines how AI is selected, implemented, and governed. It influences which problems are prioritized, which risks are tolerated, and which trade-offs are accepted. AI does not decide whether its deployment strengthens culture or weakens it. People do.

When the human layer is strong, AI is integrated with intention. Data quality improves because employees understand its importance. Governance becomes proactive rather than reactive. Teams experiment responsibly. Leaders ask not only what can be automated, but what should be.

When the human layer is weak, AI becomes transactional. Decisions default to efficiency metrics alone. Oversight becomes compliance-driven rather than principle-driven. Employees disengage, and innovation narrows to whatever produces the fastest measurable return.

In environments where employees believe AI is being introduced primarily to replace them, behavior shifts quickly. Information becomes guarded. Experimentation slows. Risk tolerance declines. People protect relevance rather than pursue improvement. Innovation narrows to what feels safe. The technology may function, but adoption becomes compliance-driven rather than curiosity-driven.

In environments where AI is introduced as a capability amplifier, the opposite occurs. Employees suggest new applications. They identify friction points leadership may not see. They collaborate across teams to refine implementation. Adoption accelerates not because it is mandated, but because it is perceived as supportive. The return on AI investment becomes not only operational, but cultural.

The technology is the same in both scenarios. The difference lies in whether the human layer feels threatened or strengthened. The difference between those two outcomes is not technical sophistication. It is human capability.

As AI becomes more accessible and more powerful, competitive advantage will not belong to the organization with the most advanced tools. It will belong to the organization that has cultivated the strongest human layer beneath them.

Tenacity, creativity, and observation cannot simply be scraped from datasets and deployed. They are developed through lived experience, failure, empathy, and resilience. The ability to read a room, sense when someone is not okay, and understand what remains unsaid are forms of intelligence that emerge from attention and presence. These capacities shape how technology is built, implemented, and governed.

Creativity is often misunderstood in the age of generative AI. When a model can produce images, essays, music, and code in seconds, it is tempting to assume creativity has been automated. But generation is not the same as creation.

AI generates by identifying patterns across vast datasets. It recombines what already exists. It predicts what is statistically likely to follow. That is powerful. It is efficient. But it is not the same as creativity born from lived experience.

Human creativity emerges from constraint, struggle, and context. It is shaped by failure, by personal history, by emotion, and by contradiction. It is the ability to connect ideas that have never previously been connected because a person has lived through something that no dataset fully captures.

In business, creativity is not just design. It is problem framing. It is the ability to redefine the question before answering it. It is recognizing that the issue presented is not the issue that matters. No model can originate that reframing without a human deciding the direction of inquiry.

At home, creativity is even more personal. It is how we resolve conflict, how we comfort someone in distress, how we adapt when life does not follow a script. These are not pattern predictions. They are responses shaped by presence and relationship.

If we outsource too much of our thinking to systems that generate quickly, we risk weakening the very muscles that allow us to create originally. And if human creativity weakens, the training data that fuels AI becomes increasingly derivative. The system scales repetition rather than innovation. Creativity, then, is not threatened by AI. It is required by it.

A human-first AI strategy begins not with code, but with people. It asks employees where friction exists and where automation could improve their work without eroding meaning or connection. It provides tools, training, and guardrails so that AI becomes a capability multiplier rather than a replacement strategy. It accepts experimentation and even failure as part of growth, recognizing that thriving in the AI shift is not about removing humans from the equation. It is about strengthening the human layer beneath the technology because when organizations invest in the depth, resilience, and creativity of their people, AI does not replace human potential. It amplifies it. And that is where both business and humanity succeed.

Thriving in the AI shift will not be determined solely by corporate strategy or technological sophistication. It will also be determined by personal discipline. Each of us makes daily decisions about how we engage with technology. We decide whether to think through a difficult conversation or allow a system to script it for us. We decide whether to sit with discomfort or seek immediate digital reassurance. We decide whether to practice creativity, observation, and connection or outsource those capacities for speed and ease.

Convenience is seductive. Friction is uncomfortable. AI removes friction efficiently. But not all friction is waste. Some friction builds resilience. Some discomfort strengthens empathy. Some effort deepens understanding.

The future of AI will not be defined by machines alone. It will be defined by the quality of the humans who build, guide, and use them. AI will be stronger, more creative, and more responsible when those humans continue to develop their own curiosity, creativity, and tenacity through meaningful human connection.

References

1. Elon University Imagining the Digital Future Center, Close Encounters of the AI Kind (Elon, NC: Elon University, March 2025).
2. Horowitz Research, *State of Media, Entertainment & Tech: Subscriptions 2025* (Irvington, NY: Horowitz Research, 2025).
3. Association for Smarter Homes & Buildings, *ASHB Smart Home Survey 2025* (ASHB, 2025).

About the Author

Brianne Bustos is the chief information officer at Ventura Community Partners, leading IT strategy, infrastructure modernization, and cross-organizational technology alignment. She drives initiatives to simplify and scale systems to support current operations and future growth, with a focus on secure, sustainable solutions that enable all business units to operate efficiently.

LinkedIn: https://www.linkedin.com/in/briedale

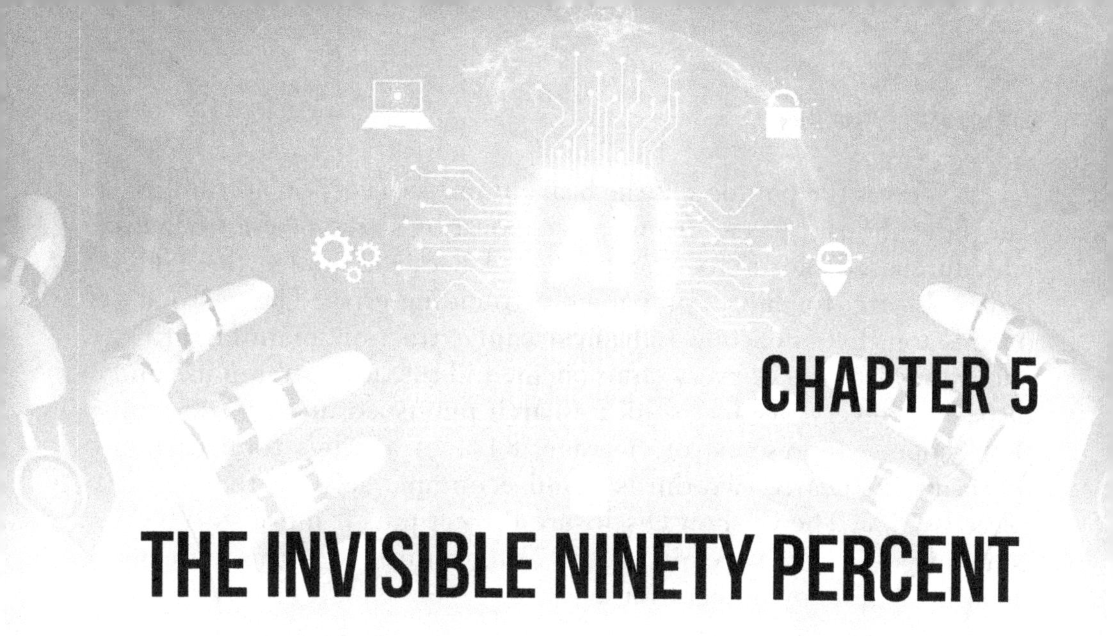

CHAPTER 5

THE INVISIBLE NINETY PERCENT

By Virbahu Nandishwar Jain
Founder & CEO, Quantisage
Hanover, New Hampshire

The greatest threat to our planet is the belief that someone else will save it.

—Robert Swan, first to walk to both the North and South Poles

Somewhere in Shenzhen, a factory is humming. It makes brushless motors: tiny, precise, the kind that spin inside your electric toothbrush. Next door, another factory stamps out lithium-ion battery cells. Across the road, a third extrudes the corrugated cardboard that will cradle the finished product on a shelf in Minneapolis. None of these facilities appear on the sustainability report of the consumer goods company whose logo adorns that toothbrush. None of their emissions are counted in the headline numbers the company displays at its annual shareholder meeting.

And yet, those factories—and the ten thousand others scattered across the supply chain—represent roughly 90% of that company's actual climate impact.

This is the paradox at the heart of modern carbon accounting. The greenhouse gases from a corporation's own smokestacks and purchased electricity—Scope 1 and Scope 2 under the GHG Protocol—are, for most companies, a rounding error. The vast mass of emissions lives in Scope 3: the upstream extraction, manufacturing, and transportation of every component, and the downstream use and disposal of every product sold. Research published in *Nature* found that Scope 3 represents, on average, 84% of a firm's total carbon footprint. In nearly two-thirds of all economic sectors, that figure exceeds 75%. The Carbon Disclosure Project puts it more starkly: A company's value chain emissions are, on average, 11.4 times greater than those of its own operations.

The numbers are not abstract. Mars, the confectionery and pet food giant, has publicly reported that Scope 3 accounts for 96% of its total greenhouse gas footprint, with the vast majority tied to agricultural raw materials. Swedish carmaker Volvo has disclosed that Scope 3 emissions linked to driving its vehicles exceed 95% of the company's total. Schneider Electric, whose CEO has called supply chain decarbonization the single largest opportunity available, estimates that reducing supplier emissions offers 20 times the abatement potential of internal operations alone. These are not outliers. They are representative of a structural reality that most corporations have spent a generation avoiding.

For decades, this invisible majority was treated as someone else's problem. Too distributed to measure. Too complex to manage. Too politically inconvenient to acknowledge. But a convergence of regulatory mandates, investor pressure, carbon border taxes, and a transformative technology called artificial intelligence has made ignorance untenable and inaction ruinously expensive.

This is a story about emissions and algorithms, but it is also a story about money, about the 40.5 billion dollars that flowed into climate technology in 2025, about how the hundred-billion-dollar market that AI-powered sustainability is projected to become within a decade, and about carbon border taxes that are turning emissions data into trade currency. It is a story about investors now recognizing that the company which builds the operating system for Scope 3 management will occupy one of the most consequential market positions of the 21st century. And most urgently, it is a story about time and about the rapidly closing window in which the choices we make will determine whether the Paris Agreement's targets remain achievable or become a historical footnote.

A Brief History of Not Counting

The GHG Protocol, published in 2001, divided corporate emissions into three scopes:

- Scope 1: direct emissions from owned sources
- Scope 2: indirect emissions from purchased energy
- Scope 3: everything else—15 categories spanning the entire value chain, from purchased goods and services to end-of-life treatment of sold products

The framework was brilliant. The execution was catastrophically lopsided. Companies raced to measure Scopes 1 and 2, because the data lived inside their own walls. Scope 3 demanded information from thousands of suppliers, many operating in developing economies with neither the tools nor the incentive to provide it. The GHG Protocol defined 15 categories within Scope 3—from purchased goods and services to capital expenditures, from upstream transportation to end-of-life treatment—divided into upstream and downstream emissions. The sheer breadth was paralyzing. Upstream emissions alone required mapping raw material extraction, supplier manufacturing processes, inbound logistics, and embedded energy, each with its own data requirements and calculation methodologies.

According to the Carbon Disclosure Project, fewer than half of the world's largest companies report Scope 3 with any rigor. Those that do overwhelmingly rely on spend-based estimates, crude calculations multiplying procurement dollars by industry-average emission factors, with error rates of up to 40%. This is, essentially, guessing and calling it compliance.

For a generation, companies measured the visible tip above the waterline and pretended the massive structure below didn't exist. Then the waterline rose.

The Regulatory Reckoning

The period between 2022 and 2026 will be remembered as the moment Scope 3 moved from the sustainability team's worry list to the CFO's balance sheet.

In Europe, the Corporate Sustainability Reporting Directive (CSRD) expanded mandatory disclosure to approximately 50 thousand companies, including non-EU firms with significant European operations. The European Sustainability Reporting Standards (ESRS) require detailed value-chain emissions data, making Scope 3 reporting a condition of market access. Simultaneously, the EU's Carbon Border Adjustment Mechanism (CBAM) introduced carbon tariffs on imports of steel, cement, aluminum, and fertilizers, forcing manufacturers worldwide to quantify the embedded emissions in goods crossing European borders. Carbon prices in the EU Emissions Trading System have breached 100 Euros per ton.

In the United States, California's SB 253 mandated that companies with revenues exceeding one billion dollars disclose their complete Scope 1, 2, and 3 emissions. The SEC proposed its own climate disclosure rules. Globally, the International Sustainability Standards Board (ISSB) released standards being adopted from Singapore to Brazil, creating a patchwork of overlapping requirements that disproportionately burdens companies without automated, multi-jurisdictional reporting infrastructure. And the Science Based Targets initiative now requires companies to include Scope 3 in near-term targets whenever it exceeds 40% of total emissions, a threshold surpassed in virtually every sector. The combined effect is that companies face not one regulatory mandate but a converging wave of them, arriving from different jurisdictions with different timelines, different reporting formats, and different enforcement mechanisms, all pointing in the same direction.

The Paris Agreement's 1.5°C pathway demands that global upstream Scope 3 emissions be reduced by 54%. We are currently on track for warming between 2.1 and 3.9°C. The human cost of this trajectory is already visible: climate-driven disruptions have reshaped supply chains in ways unthinkable a generation ago. Unprecedented heat waves shuttered factories across Southeast Asia. Drought restricted shipping through the Panama Canal. Each disruption reinforced the same message: A climate-resilient supply chain and a low-carbon supply chain are increasingly the same thing. The message to corporations is unambiguous: Measure what you have spent a generation ignoring, reduce what you have never been able to see, and do it on a regulatory timeline, or face fines, market exclusion, and investor flight.

Why This Is a Climate Tech Problem, Not a Spreadsheet Problem

A mid-sized manufacturer might source from two thousand suppliers, each generating documents in different formats, languages, and time zones. Each supplier has its own supply chain—tiers two, three, four, extending back to raw material extraction. The carbon intensity of a single component changes based on the energy grid powering the factory, the transport mode, the season, and the production batch. No spreadsheet can model this. No annual survey can capture it.

This is why Scope 3 is, at its core, a climate technology problem requiring capabilities that did not exist a decade ago: natural language processing to extract structured data from unstructured documents; machine learning to classify millions of line items against emission factor databases; graph-based modeling to represent the interconnected topology of global supply chains; real-time data ingestion from IoT sensors and energy grid APIs; and predictive analytics to simulate the carbon consequences of alternative decisions before they are made.

Within this AI revolution, three distinct transformations are unfolding. The first is visibility: seeing the full scope of value chain emissions in near real time, not through annual estimates. Natural language processing parses supplier documents—invoices, certificates of origin, energy disclosures—extracting the data points necessary for emissions calculation without requiring a single new survey. The second is intelligence: modeling the supply chain as a dynamic, interconnected graph that captures how disruption at one node cascades through the system, and how a single component's carbon intensity fluctuates with variables—energy grids, weather, commodity prices, shipping routes—that traditional tools cannot track. The third is optimization: using predictive analytics to simulate alternative suppliers, transport modes, and product designs before committing resources—transforming carbon management from backward-looking reporting into a forward-looking decision engine.

These three transformations—visibility, intelligence, optimization—are the operating system that Scope 3 has always needed but never had.

Consider the historical parallel. When electricity replaced steam in factories at the turn of the 20th century, the initial gains were modest. Manufacturers simply swapped one power source for another within the same layout. The real productivity revolution came a generation later, when engineers redesigned the factory itself around the capabilities of electric power: individual motors on each machine instead of a single driveshaft, flexible floor plans, multi-story buildings with interior workspaces no longer tethered to the window light that steam-era layouts demanded. We are at an analogous moment. Most companies today use AI to automate existing carbon accounting workflows: faster spreadsheets, smarter surveys, quicker reports. The companies that will capture the full value of this technology are those that redesign their entire supply chain decision architecture around it, embedding carbon intelligence into every sourcing decision, every logistics routing, every product design gate, and every supplier evaluation.

Carbon as the Fifth Element of Supply Chain Strategy

For half a century, supply chain strategy has optimized around four variables: cost, speed, quality, and risk. Carbon is now the fifth. This is not a metaphor. When the EU imposes a carbon border tariff, the carbon content of a shipment becomes a direct cost component. When a major retailer requires Scope 3 data as a condition of continued business, carbon compliance becomes a market access variable. When an institutional investor screens climate risk, the maturity of a company's emissions infrastructure becomes a valuation input.

The companies that embed carbon into their planning systems, vendor scorecards, and financial models will discover something the sustainability community has long argued but struggled to prove: The carbon-optimal decision and the cost-optimal decision are, far more often than executives realize, the same decision. Route optimization reduces both emissions and freight spending. Modal shifts from air to sea or road to rail reduce both CO2 and logistics cost per unit. Supplier consolidation simplifies both emissions reporting and procurement complexity. Inventory rationalization lowers both warehousing carbon and working capital. The missing piece has always been the data infrastructure to make this convergence visible and actionable in real time. Climate tech, powered by AI, is building that infrastructure.

The strategic distinction is already visible. Consider two companies in the same industry, facing the same regulatory requirements. Company A treats Scope 3 as a compliance checkbox: It hires a consultancy, produces an annual report using spend-based estimates, files it with regulators, and moves on. Company B treats carbon data as a strategic asset: It integrates real-time emissions intelligence into sourcing and logistics, renegotiates supplier contracts with carbon performance clauses, redesigns products to reduce lifecycle footprints, and qualifies for green bonds at lower interest rates. Both companies are "compliant." But Company B is building a compounding advantage that widens with every reporting cycle because carbon costs are directional. They only go up. The early infrastructure investment does not merely pay for itself; it appreciates in strategic value as the regulatory environment tightens and competitors scramble to catch up.

The investment thesis confirms the opportunity. Climate tech venture and growth capital reached 40.5 billion dollars globally in 2025. The AI-in-environmental-sustainability market is projected to reach one hundred billion dollars by 2034 at nearly 20% annual growth. Growth-stage deal counts rose 41% while seed funding declined 20%. Investors are consolidating around proven platforms with regulatory tailwinds and defensible moats, not spreading bets across unproven concepts. The World Economic Forum estimates that the climate transition requires annual global investments of at least four trillion dollars by 2030. North America dominated with 37% market share, while Asia Pacific emerged as the fastest-growing region, with India's climate tech sector projected to grow at nearly 30% annually through 2035.

For investors parsing this landscape, the signal is clear. The risk question has shifted from "Will this technology work?" to "Who will win the market?" Companies attracting capital are not science experiments. They are software businesses with recurring revenue, regulatory-mandated demand, and measurable unit economics, solving what the industry calls a "hair on fire" problem: Companies must report or face fines, and they currently have no tool to do it accurately.

The Human Layer

The most powerful AI model is useless if the procurement team does not trust its recommendations. The most granular emission estimates are meaningless if the Tier 2 supplier lacks the capability to act. The most comprehensive carbon dashboard is decorative if executives treat it as a compliance artifact rather than a decision input.

Authentic sustainability transformation requires parallel investments in change management and supplier enablement. Supply chains generate an estimated 60% of global emissions when accounting for manufacturing, extraction, logistics, product use, and disposal.

No sector, no regulation, no technology can bend the emissions curve without fundamentally transforming how they operate. Research consistently confirms the principle: When companies actively engage suppliers, those suppliers are significantly more likely to provide accurate data and take reduction actions. The World Economic Forum's Scope 3 Upstream Action Plan calls for companies to engage at least 67% of their supplier base by 2026 and for those suppliers to set science-based targets by 2028.

The old model of sending annual questionnaires is dead. Suppliers drowning in data requests from multiple customers, each in different formats, with different deadlines, using different methodologies, will not provide accurate information regardless of how many reminder emails arrive. The new model must be reciprocal. Suppliers should receive a validated carbon footprint they can use in future tenders, benchmarking data against industry peers and actionable reduction insights. When the exchange creates value for both parties, data quality follows. When it feels like a bureaucratic extraction, data quality collapses. AI augments human judgment; it does not replace the relationships, strategic context, and institutional will that drive real change. The companies that get this right will treat AI as an amplifier of organizational capacity, accelerating the visibility but never substituting for the trust that makes suppliers share real data instead of convenient data.

The Operating System the Supply Chain Has Been Missing

It is against this backdrop that I found Quantisage. The premise is born from a career spent inside the machinery of global supply chains, not theorizing about them, but living inside the data governance gaps, the supplier engagement failures, and the integration challenges that make sustainability transformation so difficult in practice. I have implemented enterprise systems that run billion-dollar supply chains. I have sat in conference rooms where sustainability teams presented carbon data that the CFO dismissed because it was 12 months old. I have watched companies spend millions on consulting engagements that produced beautiful PDF reports and zero operational change.

Quantisage exists because I became convinced that the problem was not a lack of data—the data already lives inside ERP systems, procurement platforms, and logistics networks—but a lack of infrastructure to translate that data into carbon intelligence at the speed and granularity that real decisions demand.

Quantisage is focused on building solutions in the supply chain spaces, most specifically on the climate technology dedicated entirely to one mission: making Scope 3 emissions visible, measurable, and, critically, profitable to reduce. We have built the AI Operating System for Scope 3 profitability, a platform unifying the three revolutions described in this chapter into a single, continuous decision engine. Our conviction is that carbon compliance and cost optimization are not competing objectives. They are the same objective viewed through different lenses. The route that minimizes total landed cost, inclusive of carbon liabilities, is often the route that produces the fewest emissions. The supplier portfolio that minimizes regulatory exposure is often the portfolio that maximizes procurement efficiency. The missing ingredient has always been the analytical infrastructure to prove this convergence in real time, at the granularity of individual shipments and sourcing decisions.

We have not built just a reporting tool that tells you what happened last year but a decision engine that reveals what will happen next quarter if you change nothing and what could happen if you change something specific. The distinction matters to the companies we serve and to the investors evaluating this space. The sustainability software market is littered with tools that generate beautiful reports nobody acts on. The reason is architectural: They are designed as compliance endpoints, not operational systems. They sit outside the decision workflow: consulted annually, filed with regulators, forgotten.

A platform that genuinely transforms how supply chains manage carbon must be embedded in the daily operational rhythm: in the procurement system where sourcing decisions are made, in the logistics platform where routes are planned, in the product lifecycle tool where materials are specified. It must deliver value that the supply chain team recognizes as operational value, not sustainability overhead. This is the design principle at the core of everything Quantisage builds. Carbon intelligence must become indistinguishable from operational intelligence. When the line between sustainability data and supply chain data disappears, so does the false choice between doing good and doing well.

What Comes Next

A pattern repeats across every industrial transformation: Practitioners see the problem and begin building. Regulators catch up, converting voluntary practices into mandatory requirements. Capital floods in. The transformation becomes obvious in retrospect.

We are between stages two and three. The regulations are landing. The capital is beginning to flow. The next phase—scale—will be defined not by the brilliance of any single technology but by the discipline with which organizations integrate carbon intelligence into daily supply chain decisions. The companies that build this infrastructure first will not merely grow; they will become the essential layer of a carbon-constrained global economy.

The generation now entering the C-suite is the first to have grown up with the lived experience of climate disruption: wildfires visible from childhood bedrooms, hundred-year floods occurring with annual regularity, supply chain shocks that emptied store shelves during a global pandemic. For this generation, sustainability is not an abstract policy preference. It is the baseline condition for a functioning civilization. The generation now controlling institutional capital is the first to systematically screen for climate risk. For these allocators, ESG is not a marketing label, but it is a fiduciary obligation. They are looking for companies that have internalized the reality that carbon costs are directional—always rising—and that have built the infrastructure to turn that directionality into competitive advantage rather than financial exposure.

The investment characteristics of this opportunity are distinct from earlier climate tech waves. Demand is regulatory-mandated, not voluntary. Companies must comply or face fines and market exclusion, creating a predictable, non-discretionary revenue base. The data infrastructure required is inherently sticky: Years of supplier data, customized emission models, and embedded decision workflows do not migrate easily. And we are in the narrow window between regulatory mandate and widespread compliance, the precise moment when category-defining companies are built. For investors who have watched climate tech cycle through hype and disillusionment, this moment is fundamentally different. The demand signal is not a corporate sustainability pledge that can be quietly abandoned. It is a legal obligation backed by fines, enforced by auditors, and visible to shareholders.

The invisible 90% is no longer invisible. The question is what we choose to do now that we can finally see it. The answer will come not from a single breakthrough but from the steady accumulation of better decisions, millions of them, made daily, by procurement officers choosing lower-carbon suppliers, logistics managers selecting cleaner routes, product designers specifying sustainable materials, and investors directing capital toward companies that measure what matters. AI does not make these decisions. People do. What AI provides is something that has never existed before in the history of industrial commerce: a shared, real-time, data-driven understanding of the carbon consequences of every choice across the entire value chain.

The technology that wins will not be the most sophisticated. It will be the one that makes the right decision the easy decision—ensuring every decision maker has the carbon data they need, at the moment of decision, to make the choice that is both economically rational and environmentally responsible.

That is the future Quantisage is building. That is the future the climate demands. And the story is just a beginning.

About the Author

Virbahu Nandishwar Jain built his career inside the machinery of global supply chains, spending over 20 years transforming enterprise operations across manufacturing, life sciences, energy, and the public sector on five continents. That frontline experience led him to found Quantisage, a company building proprietary AI-powered solutions for supply chain complexities and compliance, with a current focus on Scope 3 emissions management and decarbonization. A recognized voice at the intersection of AI, sustainability, and global supply chains and a sought-after speaker on supply chain issues, Virbahu has published peer-reviewed research on AI, robotics, and IoT applications in sustainable supply chains cited extensively in academic literature. He holds patents in AI and IoT for industrial manufacturing, an MBA from the Kellogg School of Management at Northwestern University, and a bachelor's in information technology. He is a chartered engineer and CPIM-certified supply chain professional.

Email: vir@quantisage.com

Website: https://quantisage.com/

Website: https://quantisage.ai/

LinkedIn: https://linkedin.com/in/virbahu

CHAPTER 6

THE 80/20 RULE OF AI TRANSFORMATION: WHY ENTERPRISE AI IS 80% PEOPLE AND 20% TECHNOLOGY

By Kyle Cousin
AI Executive
Boston, Massachusetts

Top-performing organizations follow the 10-20-70 principle: 10% of their efforts go to algorithms, 20% to data and technology, and 70% to people, processes, and cultural transformation.

—Boston Consulting Group, 2024

I've spent my career watching AI projects fail. Not because the technology didn't work. It usually did. They failed because of humans.

After leading enterprise AI transformations at a global pharmaceutical company and a private-equity-backed home services organization, I've arrived at a conclusion that may disappoint the technologists in the room: AI success is 80% people and process, 20% technology. The companies that win won't be the ones with the best algorithms. They'll be the ones who figured out how to bring their people along for the journey.

This isn't just my opinion. BCG's research confirms that 70% of AI implementation challenges stem from people and process issues, while only 10% involve the algorithms themselves. Harvard Business Review put it more bluntly: "Most firms struggle to capture real value from AI not because the technology fails—but because their people, processes, and politics do."

The two biggest pitfalls are cultural and practical. First, organizations often treat AI as a technology project rather than a transformation of how people think, work, and make decisions. Without stakeholder buy-in and a culture that values experimentation, even the best models will die on the vine. Second, too many teams try to "cure cancer" on day one, chasing moonshots before proving value. The key is to start small with initiatives that deliver measurable P&L impact. Those wins build credibility, self-fund the next wave, and gradually create momentum for scale.

In short, AI success is 80% people and process, 20% technology. It's about building belief first, then building models.

The Education: A Global Pharmaceutical Company

My first enterprise AI transformation took place inside a large, publicly traded multinational pharmaceutical company before ChatGPT existed. This context matters.

At that time, AI was dominated by classical machine learning, robotic process automation, natural language processing pipelines, and narrowly scoped automation solutions. There were no large-language-model platforms that business users could interact with directly. Everything required significant engineering, long development cycles, and deep specialization. AI literacy across the global workforce was limited. In other words, we were trying to build intelligence in a world where intelligence wasn't yet democratized.

The organization had real strengths. Its size allowed us to identify pockets of believers and gain traction in certain countries.

We had real wins. The shared services organization became early adopters, using OCR for invoice processing and expense reporting. By today's standards, these seem simple, but at the time, they were transformative. The team was bought in from top to bottom, with performance metrics directly tied to AI adoption: cost per invoice processed, cases processed per FTE, and cost savings as a percentage of the budget. We ran a global "bot-a-thon" aiming to automate one million hours in a single day. We exceeded one hundred thousand hours and called it a success, given how early we were in the journey.

But success in pockets doesn't mean success at scale. The challenges we encountered weren't unique failures. They were patterns I've since seen repeated across industries. We deployed AI rather than adopting it. There's a difference. Deployment is a technology project. Adoption is a cultural transformation. We conducted substantial training and change management activities, but employees still perceived it as learning a new tool rather than embracing a fundamentally different way of working.

No one agreed on what "good" looked like. IT optimized for accuracy. Business optimized for speed. Finance optimized for cost savings. Legal optimized for risk reduction. These conflicting metrics created perpetual misalignment and stalled adoption before it could build momentum.

The Time Savings Trap

One of the clearest systemic patterns I've seen across global enterprises is the difficulty of translating time savings into true business value. It's not a failure of any particular company. It's how large organizations are wired.

The clearest lesson came from what I now call the time savings trap. I helped lead an initiative that delivered 3.5 million hours of time savings over three years. When you first hear that number, it sounds extraordinary, the kind of metric that makes its way into board decks and investor updates.

But then came the inevitable executive question: "Okay ... so what?"

That question isn't cynical. It's rational. Hours aren't dollars, and enterprises make decisions based on P&L impact, not elapsed time. When you break it down, the math becomes almost comical: 15 minutes saved per person per day, aggregated across thousands of employees until the number sounds impressive. Individually meaningful. Collectively monumental on paper. But unless an organization changes staffing or throughput while doing roughly the same work with slightly less friction, the company continues funding large transformation budgets year after year. What started as a celebrated productivity win turned into an expensive program that never translated into a viable financial impact.

This is a universal enterprise pattern: Business cases built on theoretical time savings almost always over-promise and under-deliver, because time only turns into money when the operating model changes to capture the benefit. And capturing the benefit requires more complex conversations about process redesign, role changes, and capacity reallocation, conversations most organizations never structurally plan for.

I carried this lesson with me.

Applying the Blueprint: A Home Services Transformation

When I joined a private-equity-backed home services company focused on emergency home repair, I was tasked with accelerating digital and operational transformation by introducing scalable, modern AI capabilities. The company viewed AI as a strategic differentiator. Armed with lessons from my prior experience, we designed everything intentionally different.

First, we designed a value framework anchored in P&L impact rather than time savings. Every use case had to articulate direct cost takeout, revenue uplift, cycle-time compression, call deflection, or quality improvements tied to measurable outcomes. No more "millions of hours saved" slides, only financially meaningful, capturable value.

Second, we deliberately designed the strategy around early wins to build trust. A major lesson from my previous experience was that momentum matters. Without early wins, AI loses narrative power and political support. So we intentionally chose fast-cycle, high-visibility use cases with clear success metrics and direct customer or agent impact. This built confidence quickly and earned us the right to pursue more ambitious initiatives later.

Third, we designed the function to be business-led rather than technology-led. AI succeeds when the business drives prioritization and adoption—not IT, not data science alone. Business units co-owned the roadmap. Value cases were jointly authored. Adoption metrics were mutually accountable. This made the transformation sustainable.

The results came quickly. Call containment increased, allowing agents to focus on complex, value-added calls rather than routine transactional ones. Average handle time decreased as improvements in identity verification and triage meant agents started each call with cleaner information. Marketing performance improved. We understood not just which campaigns worked, but why, enabling predictive targeting and identifying geographic areas where future marketing would be more effective. We productionized an image fraud detection system that analyzed over one million images for fraud indicators and equipment age estimation, directly addressing margin leakage.

The 80% in Action

These wins weren't just technical achievements. They were products of deliberate cultural choices. We adopted a "value-first" mindset, not a "technology-first" mindset. AI wasn't about models, accuracy scores, or novelty. It was about business outcomes: containment, handle time, margin protection, and customer experience. Stakeholders stopped asking "What can the model do?" and started asking "What problem are we solving, and how will we measure success?"

We practiced radical transparency. We communicated openly about what was working, what wasn't yet, what the data showed, the trade-offs we were making, and the improvements coming next. This built trust early, especially with operations teams who needed to see that we spoke their language and cared about real outcomes.

We embraced small, fast, iterative wins instead of giant, perfect plans. We avoided months-long design cycles in favor of short experiments, rapid evaluations, and tight feedback loops. Teams saw progress weekly, not quarterly. Momentum became a cultural asset.

We built deep partnerships with the business. Rather than building in isolation, we embedded ourselves in every part of the business. This co-design approach ensured solutions solved real problems and created champions who advocated for expansion because they helped build it. The message was clear: AI is something we do with you, not to you.

We created accountability without blame. Clear expectations and success criteria, but also psychological safety around iteration, learning, failure, and continuous improvement. Innovation accelerated because there was no penalty for intelligent risk-taking.

Finally, we celebrated wins, even small ones, to reinforce belief. Containment gains, triage improvements, fraud catches, and marketing prediction accuracy improvements. This helped the organization believe that AI wasn't just a future promise. It was happening now and working.

Why Leadership Made It Possible

None of this would have worked without executive support, and I don't mean passive approval. Leadership established AI as a core pillar of the company's future, not an experimental side project. They gave the program executive sponsorship, a defined remit, and a seat at strategic discussions. This signaled to the organization that AI was foundational to how we would operate going forward.

Leaders aligned behind business-first outcomes, consistently reinforcing that AI must deliver measurable operational and financial impact. This eliminated distractions, focused prioritization, and gave us cover to deprioritize work that wasn't meaningfully tied to outcomes. They embraced fast iteration over perfection, understanding that early versions wouldn't be perfect, and that was the point.

Most importantly, they protected the work from unnecessary friction. They removed political noise, competing priorities, and organizational blockers that otherwise slow AI programs to a halt. They created the permission structure for business leaders to co-own AI outcomes.

What Enterprise Leaders Must Understand

If there's one argument I'd make to anyone leading an AI initiative, it's this: AI only creates enterprise value when leaders treat it as a business transformation, not a technology initiative, anchored in measurable outcomes, empowered teams, and a culture willing to learn and adapt.

Leaders often misunderstand that AI is not a tool you install, but a capability you build. Its success depends far more on operating models, processes, and people than on algorithms or vendor platforms. Most failures come from organizational friction, misalignment, and unclear ownership, not from the technology itself.

Leaders routinely undervalue the importance of culture, change management, and early wins. They overlook the power of small, fast, high-credibility use cases. They underestimate how crucial it is to measure value in P&L terms rather than theoretical time savings. They skip foundational investments in governance, data access, and operating models that enable scale.

The first step is alignment: Define the business outcomes AI must deliver, assign clear ownership, and establish how success will be measured. Only after this should you invest in platforms or models.

The mindset shift required is profound. Enterprises must move from "AI as a project" to "AI as a capability and a new way of operating." This means adopting a mindset of continuous learning, iterative improvement, shared ownership between business and technology, and a willingness to redesign processes, not just automate them. Scaling AI requires embracing progress over perfection, outcomes over activity, and culture over code.

The technology will continue to evolve faster than most operating models, cultures, and governance structures. The lessons from the pre-ChatGPT era remain as relevant today as they were then. Build belief first. Then build models.

About the Author

Kyle Cousin is an SVP of AI with a career spanning biotech, pharmaceuticals, medical technology, and home services. He has led organizations through the adoption, scaling, and operationalization of AI, from early machine learning programs in the pre-ChatGPT era to modern enterprise LLM deployments and data monetization strategies. His accomplishments include building global AI organizations, delivering double-digit reductions in operating costs, protecting margins through fraud detection, and generating new revenue streams through predictive data products. Kyle is recognized for bridging vision and execution, an AI strategist who defines the future state and an operator who delivers it.

LinkedIn: https://www.linkedin.com/in/kcousin/

CHAPTER 7

UNIFYING AI GOVERNANCE AND SECURITY: A COMPREHENSIVE FRAMEWORK FOR MANAGING AI RISK

By Ervin Daniels
Sr. Data & AI Platform Technical Specialist
Atlanta, Georgia

Everyone's chasing Nvidia, but IBM is quietly making real money from AI.

—Investors Observer, LinkedIn post

The Benefits and Rise of AI

Artificial intelligence (AI) is transforming how organizations operate, innovate, and compete. Adoption continues to accelerate as businesses become more aware of its potential. One of the most disruptive advancements is generative AI (GenAI), which creates new content, text, code, images, and more from simple prompts. Its impact spans financial services, healthcare, marketing, entertainment, and other sectors.

The explosion of ChatGPT brought AI into global public shockwaves. Almost overnight, it demonstrated its extraordinary AI power. Businesses now use GenAI not as a novelty but as a strategic asset to automate tasks, improve decision-making, streamline operations, and enhance customer experiences. As adoption continues to expand, AI is becoming a competitive differentiator and a core driver of business value.

The Emerging Landscape of AI Risk

AI is highly powerful and beneficial, but it also introduces new risks for organizations. AI systems, especially GenAI, can produce incorrect or inconsistent outputs, make flawed decisions, or expose an organization to reputational or business harm. These risks generally fall into the business, security, and privacy categories. Organizations can address them most effectively when AI governance and security practices work together. Although the two domains overlap in their goals, they each address different aspects of AI risk, and most organizations today have weak or nonexistent AI governance structures.

Business Risk and Organizational Ownership

Once organizations understand that governance and security can mitigate risk, they must better understand the risks and shift their focus to the underlying issues that drive them. Managing AI risk requires clarity about who is accountable for what. In most organizations, the ultimate decision-making authority is likely the chief risk officer (CRO), who oversees decisions on all organizational risks, including AI system development and deployment; however, effective AI governance relies on a shared accountability model, where executive leaders set the organization's risk tolerance and oversight while operational teams, depending on the different organizational structure, manage risks tied to their stage of the AI lifecycle.

From a security perspective, the chief information security officer (CISO) faces the challenge of identifying generative AI use and implementing security controls centrally as departments accelerate the adoption and development of generative AI systems. The CISO's focus is on protecting AI systems from internal risk and external attacks. Their responsibilities include preventing unauthorized access, monitoring for threats, reducing vulnerabilities, and safeguarding AI models, applications, and the data they rely on. Or there might be a security lead who reports to the AI innovation executive, responsible for securely preparing data and models across the various stages of the generative AI development lifecycle.

These roles are also concerned with the problem of shadow AI. Shadow AI refers to rogue artificial intelligence (AI) systems or applications implemented and used within an organization's environment without the knowledge or approval of the central IT, security, or governance teams. Shadow AI poses an even greater risk because it's easy for anyone to deploy.

Understanding the Root Causes of AI Risk

As organizations adopt more AI models, several core risks begin to surface, each carrying potential regulatory, reputational, and operational consequences. Understanding these origins is essential for organizations looking to implement effective risk-mitigation strategies and build trustworthy AI systems. These risks fall into three main categories: input and inference risks, output risks, and non-technical risks.

Input and inference risks start with the data used to train AI. Poor-quality or biased data can introduce unfair outcomes, privacy issues, or misinformation. During inference, when the model generates predictions, these weaknesses can surface as incorrect, harmful, or misleading outputs.

In terms of output risks: AI systems may produce toxic, biased, or fabricated content, including misinformation or hallucinated answers that appear confident but are false. These outputs can damage trust and create compliance or safety issues.

Because not all risks are technical, there's also the possibility of non-technical risks. Under this umbrella lie fairness, transparency, accountability, and the broader societal impact of AI, including how decisions affect people and jobs.

Understanding the root cause of AI risk is like inspecting the foundation of a house before building the upper floors. If the foundation is cracked, that means data is poor, oversight is weak, or systems are vulnerable, so everything built on top becomes unstable. But when the foundation is solid, the entire structure becomes safer, more reliable, and able to withstand pressure. The same is true for AI: Strong governance and security at the foundation prevent minor issues from becoming structural failures later.

The Regulatory and Compliance Landscape

AI regulations are emerging rapidly worldwide, and the landscape is still evolving. There is no single, overarching federal law governing AI in the United States, but a complex patchwork of rules and regulations is emerging globally and at the federal and state levels. While dozens of frameworks exist globally, a few are shaping how organizations approach governance today, including the EU AI Act, an established risk-based framework to ensure AI systems in Europe are safe, the US NIST AI Risk Management Framework, and the AI regulatory and compliance guide. Auditors now require proof of responsible AI practices. They may review training documentation, data lineage, results of fairness and bias testing, explainability records, and evidence of human oversight. Missing controls may lead to failed audits, fines, litigation, and forced model shutdowns. Think of these AI rules like seatbelt laws for technology. Just as seatbelts were introduced to make cars safer, these AI regulations aim to protect people and businesses from potential risks posed by AI.

Governance ensures AI behaves ethically, fairly, and consistently, but it cannot prevent external manipulation or internal misuse. However, AI introduces a new attack surface that traditional security approaches cannot fully protect.

The next step in understanding AI risk is understanding how external actors or attackers exploit AI systems and why AI requires security measures beyond what organizations typically use. Even well-governed AI systems are vulnerable to attacks designed to manipulate outputs, corrupt training data, steal sensitive information, degrade performance, or disrupt availability.

AI Security: Protecting the GenAI Models, Data, and Usage

As organizations develop GenAI and deploy more systems into production environments, they become susceptible to cybersecurity attacks. AI security is a capability that protects the models, applications, data pipelines, and operational workflows from these threats. A CISO's responsibility in ensuring confidentiality, integrity, and availability (the CIA triad) extends directly to AI.

There are six main ways attackers might try to undermine these CIA principles in AI systems. Each one has a special way of causing trouble. One primary attack is prompt injection, which is like the AI version of a SQL injection attack, meaning attackers try to trick the system with harmful instructions that lead to incorrect or misleading outputs. Another risk is infection: AI models downloaded from open-source sites may come preloaded with malware, creating a supply-chain problem. Evasion attacks entail attackers tweaking inputs just enough to fool the AI into making the wrong decision. Poisoning is another threat in which attackers tamper with the training data so the AI learns the bad patterns. Extraction attacks focus on stealing valuable information, such as sensitive data or intellectual property, by posing a series of questions that gradually reveal protected content. Finally, denial-of-service attacks overwhelm the AI system with excessive requests, slowing it down or causing it to crash.

Think of securing AI as running an airport. For example, airport security must check every bag, and every passenger and flight path must be checked. Attackers may try to smuggle harmful items in, bypass checkpoints, overload the gates, or disguise their intent. AI security functions as the airport's screening, monitoring, and control systems, ensuring that everything entering, exiting, and moving through the system is safe, verified, and trustworthy.

Challenges to Securing and Governing AI Deployments

As organizations deploy more AI systems, they encounter new and evolving challenges. Many teams implement AI across multiple hybrid cloud platforms and environments without a centralized inventory, making it challenging to gain visibility into which models exist and where they run, thereby increasing the risk of shadow AI. AI pipelines under development can unintentionally expose sensitive data, increasing the risk of data leaks. Regulations add another layer of complexity. As policies evolve, organizations must constantly reassess whether their AI systems still meet the latest regulatory and compliance requirements. Maintaining model quality is also challenging, particularly when model drift occurs, which is the degradation of an AI model's performance over time as the real-world data it encounters changes from the data it was trained on initially. Operational misconfigurations, such as overly permissive access settings, introduce avoidable risks, such as unauthorized or excessive privileged access.

As organizations deploy more AI, governance becomes increasingly complex. What works for a single model rarely works for hundreds. These challenges make it essential for governance and security to operate as a unified effort throughout the AI lifecycle.

A Unified Framework for AI Governance and AI Security

As we've discussed, governing and securing AI cannot be treated as separate efforts. AI introduces new ethical, regulatory, and operational responsibilities while simultaneously expanding the organization's attack surface in ways traditional security models were never designed to handle. To manage AI risk safely and responsibly, organizations need a unified framework that brings together people, processes, and tools across both governance and security disciplines.

Two categories of tools and capabilities typically support this integrated approach. On one side are AI security platform solutions:

- AI discoverability: This platform automatically discovers AI models known and unknown, gaining visibility across organizational use through data, models, and applications.
- Security posture management: This platform detects and assesses security vulnerabilities, prioritizes risks, and recommends actions.
- Data security: It protects sensitive data from unauthorized access and ensures training and fine-tuning of data.
- Operations: It uncovers potential cybersecurity threats to mitigate operational disruptions in a unified workflow.
- Monitoring: This platform continuously monitors AI activity for misuse and abnormal behavior.

These capabilities strengthen the technical foundation of AI by ensuring that models remain protected, resilient, and safeguarded against threats.

On the other side are AI governance platforms solutions that focus on:

- Automating the capture and documentation of model facts throughout the AI lifecycle
- Evaluating metrics for several use cases, including text summarization, classification, translation, retrieval, augmented generation, and Q&A
- Validating prompt performance to ensure accurate output and avoid producing potentially harmful or inappropriate content

- Monitoring AI performance and setting pre-defined thresholds for both inputs and outputs to generate proactive alerts
- Facilitating compliance management of requirements and controls to support internal policies and procedures, industry standards, and regulations

When these two domains work independently, gaps inevitably emerge. Security without governance protects the system, but not the decisions organizations make strategically or tactually. Governance without security ensures responsible use but cannot defend against manipulation or attack.

Bringing It All Together

Only when combined do they create a complete lifecycle-wide approach to AI risk. Only a unified approach can deliver a comprehensive strategy for mitigating AI risk. By combining an AI governance platform with an AI security posture management tool, organizations gain a single view and an integrated framework that secures data, models, and usage while ensuring compliance, transparency, and responsible operations.

A unified AI governance and security platform brings everything together into a single, coordinated system for managing AI risk. It begins with a unified AI inventory that gives both business and security teams a clear view of every AI system, what it does, where it runs, and who owns it. On top of this visibility, the combined platform enables proactive risk and compliance management, allowing teams to prioritize and monitor model health, performance, policy compliance, and security vulnerabilities before they escalate. Governance contributes to the documentation, oversight, explainability, and policy enforcement needed to ensure responsible use. At the same time, security capabilities provide the auto discovery, protection, monitoring, and vulnerability management required to safeguard models, data, and usage.

When these capabilities converge into a shared workflow with a unified dashboard, teams gain full lifecycle context on how each model was developed, how it is behaving, where weaknesses exist, and whether it meets regulatory and ethical expectations.

Integrated operations ensure that product, risk, compliance, and security stakeholders work from the same information and processes, enabling consistent management across the enterprise. Together, this unified approach forms a comprehensive framework that empowers organizations to innovate responsibly while defending against the operational, regulatory, ethical, and security risks that accompany modern AI.

Conclusion

AI is transforming how organizations operate, make decisions, and deliver value, but it also introduces new risks that require thoughtful, coordinated management. By bringing AI governance and AI security into a single, integrated framework, organizations can ensure their AI systems are not only innovative and effective but also safe, compliant, ethical, and resilient. The future of AI belongs to those who can balance progress with protection, building systems that are both trusted and secure, from design to deployment and beyond.

About the Author

Ervin Daniels is a seasoned cybersecurity leader with extensive experience in information technology, cybersecurity, technical architecture, and hybrid cloud environments. As a cybersecurity technical specialist at IBM, he focuses on data security, AI security, and quantum-safe readiness for some of the world's largest financial institutions. A passionate advocate for trustworthy and responsible AI, Ervin writes and speaks frequently on AI governance, cybersecurity resilience, and emerging technology risks. He is an active mentor and a frequent speaker, committed to advancing cybersecurity education and preparing professionals for the next wave of technological disruption.

Ervin also serves on his alma mater's alumni board, contributing to the development of future-ready cybersecurity and fintech programs. He holds an MBA in information security and maintains industry certifications, including CISSP, CCSP, and CBCP.

Email: ervin.daniels@ibm.com

Website: https://ervindaniels.com

CHAPTER 8

THE HIDDEN SOCIAL SYSTEM BEHIND EVERY AI SYSTEM

By Dr. Cecilia Dones, DBA
Professor, Founder, Chief Data Officer
New York, New York

We become what we behold.
We shape our tools and then our tools shape us.
—Father John Culkin (commonly attributed to Marshall McLuhan)

Ask a data scientist what their model tells them and you will hear probabilities: confidence intervals, precision scores, calibration curves, error bands. The language is careful. A 73% likelihood is not a guarantee. A forecast is not a promise.

Ask an executive what the AI tells them and the language often shifts. The output becomes directional. It tells us what to prioritize, where to allocate, which segment to target, which risk to mitigate. The caveats soften. The distribution compresses into a recommendation.

Somewhere between the model and the meeting, uncertainty disappears. This disappearance is structural. Organizations are built to move. Boards expect direction. Investors expect confidence. Teams expect clarity. In that environment, probabilistic nuance can feel like drag. A range requires interpretation; a single number feels decisive.

The core challenge of AI in organizations, then, is not only technical adequacy. It is interpretive discipline. Every AI system produces outputs that are equivocal and open to multiple plausible readings. Whether those outputs become insight or illusion depends on how the organization makes sense of them.

Karl Weick's *Sensemaking in Organizations* (1995) provides a useful lens. Weick argued that organizations are not primarily information-processing machines but sensemaking systems. Faced with ambiguity, people extract cues, construct plausible narratives, and act. Meaning is not discovered; it is enacted. Importantly, sensemaking is grounded in identity. Individuals interpret events in ways that preserve coherence with who they understand themselves to be.

AI does not eliminate equivocality. It translates it into numbers. It offers probabilities instead of impressions. But it still requires interpretation. And interpretation is social.

The question is not simply "How accurate is the model?" It is "How does our organization understand the ambiguity the model presents?"

What We Thought We Were Building

The early promise of enterprise AI was elegant. Replace human inconsistency with algorithmic objectivity. Replace intuition with data. Replace negotiation with neutral computation. More data plus smarter models would yield better decisions.

There is truth in this aspiration. Human judgment is biased and noisy. Standardization improves performance in many domains.

In forecasting, pricing, logistics, and fraud detection, disciplined modeling frequently outperforms unaided intuition.

Vendors amplified the promise. Dashboards offered a "single source of truth." Platforms promised "360-degree visibility." Machine learning would "eliminate bias" and "optimize outcomes." The implication was subtle but powerful: Once we have the right model, the decision becomes self-evident.

What this narrative overlooked was that organizations are not neutral containers waiting for better data. They are structured systems of authority, incentives, and norms. As Weick observed, people do not passively receive information; they enact environments through interpretation. Introducing AI does not bypass this process. It enters into it. In many cases, the technology performs exactly as designed. The surrounding system adapts around it.

When the System Changes Around the Algorithm

Consider a national retailer that implemented a machine learning demand forecasting system. The model was technically robust. In pilot markets, it reduced both overstock and stockouts, improving margin performance and inventory efficiency. It is optimized for expected revenue under defined constraints.

A regional director, analytically capable and operationally seasoned, embraced the system early. For several months she followed its recommendations closely. Results improved. Then reporting dynamics shifted. Senior leadership began reviewing store performance against model forecasts. Variance required explanation. Stores that exceeded projections were praised for disciplined execution. Misses attracted scrutiny.

In executive reviews, the director found herself explaining why certain stores underperformed forecasts. She attempted to articulate what the documentation made clear: The model optimized for probability, not certainty. Local competitor promotions, weather events, and supply chain delays introduced variability beyond its parameters. The nuance did not travel well. In performance forums, variance looked like failure. Peers who consistently "beat the forecast" gained reputational capital.

One colleague confided that he had learned to interpret what the system was "really saying." In practice, this meant selectively adjusting recommended allocations based on contextual knowledge. He did not describe these adjustments as overrides. He framed them as refinements. He did not challenge the model publicly. That would have signaled resistance to the company's AI strategy.

Over time, informal calibration practices spread. Store managers quietly hedged certain recommendations. Regional teams developed heuristics about when to trust the model and when to temper it. Officially, the organization was data-driven. In practice, it operated through a parallel layer of undocumented human judgment. The AI continued producing forecasts. Meetings continued referencing its outputs. But actual decision authority had shifted into the interpretation layer.

The director described a growing fatigue. She spent more time managing how AI outputs were discussed than improving operations. She began to doubt her own instincts. When she sensed misalignment between local reality and model output, she questioned whether her skepticism reflected bias or insight. As Weick would predict, her identity as a competent operator was now entangled with the presence of the model. Who was she, professionally, if the algorithm knew better? The technology had not failed. The sensemaking system surrounding it had evolved in ways no one explicitly designed.

The retailer described here is a composite. No single firm contains this story in its entirety. Yet elements of it appear, in fragments, across organizations adopting predictive systems at scale: the quiet override, the performance pressure, the public alignment and private adjustment. That is precisely the point. These dynamics are not isolated anecdotes. They are structural tendencies that emerge when technical deployment outpaces interpretive design.

Authority, Incentives, and Culture: The Architecture of Interpretation

Weick emphasized that sensemaking is social, ongoing, and grounded in identity. People ask, implicitly, "What is happening?" and "Who am I in relation to it?" These questions intensify in AI-enabled environments.

Three structural forces shape how AI outputs are interpreted: authority, incentives, and culture.

Authority: Who Gets to Define Meaning?

Every organization has implicit hierarchies of epistemic legitimacy. Junior analysts may detect anomalies first, but raising concerns about a model's limitations can feel risky. Mid-level managers may perceive contextual gaps yet lack political cover to challenge executive-endorsed systems. Senior leaders, meanwhile, may be removed from operational nuance.

The AI's output is always visible; it produces artifacts. Human dissent must navigate hierarchy.

A useful diagnostic question for leadership teams is simple: Who can publicly state that the model's recommendation is flawed without reputational cost? If the answer is unclear, interpretation risk rises. Silence replaces scrutiny not because employees are complacent, but because the cost of speaking exceeds the perceived benefit. Introducing AI does not remove politics from decision-making. It relocates politics into interpretation, into who gets to define what the numbers mean.

Incentives: What Gets Rewarded

Incentive structures rarely forbid questioning AI. Instead, they reward visible alignment. Performance reviews may highlight AI adoption metrics. Career advancement may favor leaders who champion automation initiatives. Budgets may prioritize teams demonstrating measurable "AI leverage."

When adoption becomes conflated with effectiveness, decision quality can recede. In such environments, it may feel safer to attribute outcomes to "the model" than to surface nuanced judgment calls. Accountability becomes less about reasoning and more about alignment.

If results are favorable, the organization celebrates data-driven success. If outcomes disappoint, discussion may shift toward recalibration or data gaps rather than toward the interpretive process itself. Responsibility does not vanish, but its boundaries blur.

Culture: What Counts as Knowledge?

Culture shapes what evidence is considered credible. In AI-enabled environments, quantitative outputs can acquire symbolic authority. A dashboard visualization can carry more weight than years of experiential knowledge.

This shift can correct historical overreliance on anecdotes. It can also narrow the aperture of what is considered relevant. Local expertise often captures variables models cannot easily encode: regulatory nuance, emerging customer behavior, operational fragility.

When the dominant narrative becomes "the AI knows," experiential insight may retreat into informal conversations rather than structured deliberation. People still interpret; they simply do so offstage. Authority, incentives, and culture together form a sensemaking infrastructure. This infrastructure determines how equivocal information is stabilized into action. As Weick argued, organizations enact the environments they believe they inhabit. If the environment is framed as numerically definitive, questioning can feel like regression rather than rigor.

Quiet Shifts in Organizational Life

The most consequential changes associated with AI adoption are often subtle. Expertise may be redefined. Influence can shift toward those fluent in model interfaces and away from those grounded in contextual depth. This can enhance analytical discipline. It can also compress the definition of competence.

Accountability structures may evolve. If "the model recommended it" becomes an adequate explanation, ownership of outcomes can appear distributed between human and system. Without explicit norms, it becomes unclear who is responsible for interrogating assumptions.

Risk tolerance can recalibrate. Numerical probabilities create an impression of precision that feels more stable than it is. Decisions may be justified through statistical framing rather than through explicit articulation of trade-offs.

Professional identity can shift as well. When individuals perceive that their judgment is systematically discounted, engagement can decline. This is not resistance to analytics; it is uncertainty about where one's expertise meaningfully contributes.

None of these shifts are inevitable. They are patterns that emerge when organizations adopt powerful predictive tools without equal attention to how those tools are interpreted.

Designing for Better Sensemaking

If AI introduces equivocal information into decision environments, strengthening interpretation requires deliberate design.

In executive forums, move from asking "What does the AI recommend?" to ask "What objective is the model optimizing?" and "Under what conditions would this recommendation fail?" These questions foreground assumptions rather than conclusions.

When leaders override model outputs, capture the rationale and outcome. Treat overrides not as defiance but as structured information. Over time, this builds institutional memory about boundary conditions and contextual variables.

Decouple adoption metrics from decision quality. Utilization matters. But high utilization does not guarantee thoughtful interpretation. Evaluate decisions based on outcomes and learning loops, not merely on alignment with algorithmic output.

Finally, normalize constructive dissent. For consequential decisions, create structured space to articulate disconfirming scenarios. Ask explicitly what would need to be true for the recommendation to be wrong. If no plausible answer emerges, the organization may not have eliminated uncertainty; it may have suppressed it. These practices are pro-interpretation. They acknowledge that sensemaking is ongoing and that models, however sophisticated, do not exempt organizations from that work.

Returning to the Director

Two years after the initial rollout, the retailer revised its planning process. Cross-functional teams reviewed major forecasts together. Model documentation was discussed explicitly. Overrides were recorded and tracked. Performance recognition emphasized decision rigor and learning, not just forecast adherence.

In one meeting, the AI projected unusually strong demand in a particular district. The director felt uneasy. Local competitor activity suggested saturation. Instead of quietly adjusting allocations, she articulated her concern in the forum. The team examined the model's assumptions and identified a lagging external variable that had skewed recent data. They adjusted the forecast transparently and documented the rationale. Uncertainty did not disappear. It became discussable.

What changed was not the algorithm. It was the social architecture around it. The director no longer felt compelled to choose between loyalty to the model and loyalty to her experience. The organization had not abandoned AI. It had integrated it into a more deliberate sensemaking system.

The Organizational Question

Weick described sensemaking as the ongoing structuring of the unknown. AI expands the domain of the unknown while presenting it in quantified form. The temptation is to treat numerical expression as closure. However, probabilities remain probabilities.

Every AI system operates within a social system. That system determines whether uncertainty is compressed into premature certainty or examined with discipline; whether dissent is muted or legitimized; whether accountability is clarified or blurred. The strategic question for leaders is not only what their models can predict. It is what kind of interpretive system they are building around those predictions.

Technology will continue to advance. Models will grow more powerful. But the interpretive burden will remain human. As long as organizations consist of people navigating ambiguity together, sensemaking will sit at the center of decision quality. The challenge is not to eliminate uncertainty. It is to build organizations capable of living with it intelligently, and to ensure that when the data leaves room for doubt, the conversation does not close prematurely.

References

Weick, Karl E. *Sensemaking in Organizations*. Thousand Oaks, CA: SAGE Publications, 1995.

About the Author

Dr. Cecilia Dones is an AI and analytics practitioner-academic, recognized international speaker, and founder of 3 Standard Deviations. Her work sits at the intersection of technology, data, and trust, helping organizations translate complex AI and analytics into clearer, more human decision-making. She has held leadership roles within Fortune 500 companies, including L'Oréal and LVMH, and has taught at Columbia Business School and MIT Sloan.

Cecilia studies how technology mediates lived experience and how data systems shape relationships, leadership, and organizational culture. Her doctoral research focused on interpersonal trust in digital environments, bridging behavioral science and analytics. She writes widely on these themes in her newsletters "Authentic Interactions" and "What Is a CMO to Do with AI?"

Through advisory work, executive workshops, and keynote speaking, she equips leaders to navigate AI with strategic clarity, intellectual rigor, and disciplined sensemaking.

Email: cecilia.dones@3standarddeviations.com

Substack: https://www.authenticinteractions.substack.com

LinkedIn: https://www.linkedin.com/in/ceciliadones

CHAPTER 9

LEADING THROUGH THE SHIFT: HOW MODERN EXECUTIVES NAVIGATE AI-DRIVEN TRANSFORMATION

By Steffen J. Ehl
Business Executive and Advisory Board Member
Düsseldorf, Germany

The greatest danger in times of turbulence is not the turbulence itself, but to act with yesterday's logic.

—Peter Drucker

AI first hit most leadership agendas as noise, not strategy. A board member forwards a headline: "Companies That Don't Embrace AI Will Die." A customer asks how your product compares to a new "AI-powered" competitor. At the same time, regulators and lawyers start asking uncomfortable questions about data, privacy, and responsibility. Benefits and risks become so significant that engaging with AI is important for every leader, and that is probably the reason why you read this chapter.

In this environment, many executives feel the same tension: "We cannot ignore AI, but we also do not have the time or energy to become AI experts."

My own journey has been different. Having worked in the tech space for companies like Cisco, Oracle, and Synology, "AI" has been part of my world for many years, especially used in enterprise contexts. Technology helped industrial clients answer questions like: "Which part of a production machine is most likely to fail next?" Predictive maintenance enabled them to repair equipment before it actually broke.

Public sector clients asked questions like: "Can we detect early signs of risky situations in a crowd?" By analyzing video footage and flagging unusual behavior patterns, security teams could intervene before an incident escalated. Projects like this also illustrated how important it is to think carefully about ethics, bias, and clear guardrails when deploying AI in sensitive contexts.

From this perspective, one thing is very clear: The technology is not the bottleneck anymore. The real challenge is leadership: how executives set direction, define responsibility, and shape a culture that can work confidently with these tools.

AI forces leaders to make decisions about responsibility, risk, and culture faster and with less certainty than they are used to. It exposes whether an organization has clarity about its data, processes, and values because AI amplifies both strengths and weaknesses.

To help their organizations win a competitive advantage and create real, sustainable value through the AI shift, executives don't need to deep-dive into technology, but they must do the following:

- Take clear ownership of AI and its responsible use and keep it as a regular topic on the agenda

- Invest in people and culture so that teams build the skills and confidence to work with AI
- Experiment quickly and safely with low-risk, high-impact use cases

This chapter will unpack these three responsibilities and turn them into a practical playbook for modern executives.

Responsibility 1: Ownership and Agenda: Putting AI Where It Belongs: At the Top

To see how much priority AI really has in an organization, you do not need to see their strategy paper. You only need to look at the org chart. In many organizations, AI sits too low and too isolated to make a substantial difference.

Often it lives somewhere between IT, innovation, and "someone in marketing who is good with tools." That might be enough for a few pilots, but it is not enough for real, sustainable value with potential to disrupt the organization. If AI affects strategy, customers, employees, and reputation, it cannot remain a side project. It must become an executive leadership responsibility.

A helpful way to think about the role of AI in an organization is to compare it with finance. Both are cross-functional topics and touch every product, every department, every decision. You would never delegate finance entirely to a single team below the leadership level and then ignore it at the top. There is always someone on the executive team who owns finance, and the whole leadership team is expected to understand the basics and make decisions with financial impact in mind.

AI's place in the organization is very similar. It is a cross-functional responsibility. It touches how you design products, serve customers, run operations, treat data, and manage risk. You cannot simply "park" AI in IT or innovation and be done with it. One member of the executive team needs to own it, just as one member owns finance. That person needs a clear mandate to shape how AI is used across the whole organization.

At the same time AI must be understood across the organization. People across all functions need a basic level of AI literacy and clarity about what is expected from them. That is why AI needs to become a continuous topic on the agenda in the regular conversations leaders have with their teams.

We can already see this logic becoming reality in the public sector in the United States, where federal agencies have been required to designate a senior leader, often called a "chief AI officer," who is responsible for coordinating AI strategy, governance, and risk management across the agency. The principle is the same: There is a named person at the top who is accountable for how AI is used.

Some CEOs are making this point very explicitly like Sanofi's CEO Paul Hudson: "AI may be the greatest revolution since the printing press. If you're planning to delegate the AI revolution, then good luck to you." He argues that senior leaders must personally champion AI across teams and remove roadblocks for innovators, rather than pushing responsibility down the org chart. AI is not a specialist hobby; it is part of the CEO's job.

In my own experience, the difference between AI projects that quietly die and those that create real impact is rarely a specific algorithm. It is whether an executive has publicly taken ownership, linked AI to the core mission, and made it a standing item in leadership conversations. Once AI is anchored at that level, technical teams can move faster and more safely because they know the direction, the expectations, and the limits.

Responsibility 2: People and Culture: Building Skills and Confidence, Not Fear

Once AI has an owner at the top, the next question is about the organization's culture. For many employees, AI first shows up as a threat. It sounds like job cuts, surveillance, and yet another wave of tools they did not ask for. If leaders do not address this honestly, fear will quietly undermine every AI initiative. Employees' fear of AI is not a side issue; it is a substantial problem. Everyone has heard that AI will take over work and kill jobs. Who wants to collaborate with someone who might replace them?

This affects management layers as well. In the past, some spent time aggregating data, tweaking numbers, and preparing slides for higher levels. Today the CEO and the trainee can look at the same dashboards and the same real-time data.

Simon Sinek offers a useful reminder in the middle of all the technological excitement: No amount of innovation can replace the basics like kindness, respect, and real connection. The "rules of humanity" are timeless. Trust, clarity, and a sense of belonging become more important in the age of AI. If leaders want their teams to experiment with AI, they need to create a kind of "we've got your back" culture so that people still experience basic human decency and care in the way decisions are made and communicated.

Another part of investing in people is building skills. Almost everyone will need a basic level of AI literacy: understanding what current AI can and cannot do, how to frame good questions, how to check outputs, and how to recognize risks. In *Human + Machine*, Paul Daugherty and H. James Wilson describe the age of "human-machine teamwork" and argue that AI is about giving employees new powers when work is redesigned around collaboration between humans and machines.

They introduce the idea of "fusion skills": capabilities that combine human judgment, empathy, and creativity with machine speed and pattern recognition. Companies that benefit most from AI treat it as an investment in human talent first and technology second, redesigning processes so that people and machines each do what they do best.

Pat Gelsinger, former CEO of Intel, often argues that technology itself is neutral—it can be used for good or for harm—and that it is the job of technological leaders to shape it as a force for good. When leaders introduce AI into their organizations, they are not just changing processes; they are making ethical choices about how people will be treated, which tasks will be automated, and where human judgment remains central. Seeing technology as neutral but powerful pushes leaders to ask: How do we want AI to change the daily experience of our employees and customers?

Practically, investing in people and culture can look very ordinary and concrete:

- Offer simple AI learning paths tailored to different roles.
- Create internal communities or "AI circles" where colleagues discuss use cases and concerns.
- Model learning by visibly experimenting with AI in front of your team instead of pretending to have all the answers.
- Celebrate stories where AI made someone's work more meaningful or safer, not just faster.

When people feel informed, equipped, and supported, AI becomes an opportunity they can lean into, not a wave they hope will somehow pass them by.

Responsibility 3: Experiments That Matter: Low-Risk, High-Impact Use Cases

Once AI has clear ownership at the top and people begin to build skills and confidence, the next question becomes very practical: Where do we start? Many organizations get stuck between fear and hype. They either wait for the "perfect" masterplan, or they launch unconnected pilots that end up being irrelevant for real impact. In both cases, AI remains an idea rather than a source of value.

The best way is through a small number of well-chosen experiments that are low-risk and high-impact. Low-risk means they are contained, reversible, and run with a human in the loop. High-impact means they are close to the core mission of the organization and clearly linked to measurable outcomes.

Daugherty and Wilson's framework in *Human + Machine* is helpful here. To summarize their approach: They highlight the need for guardrails and human checkpoints: places in the workflow where people can monitor, understand, and, if necessary, override AI decisions so that automation stays aligned with organizational values and avoids harmful outcomes. For example, a bank's loan-approval system might let an AI model pre-score applications, but a human credit officer reviews borderline cases, can override the recommendation, and regularly checks for bias in the outcomes.

In modern AI architectures, this idea of guardrails is implemented with an additional review layer. One model generates an answer for the customer, while a second model, or a dedicated "safety layer," reviews that answer before it is sent out. This reviewer checks for policy violations, sensitive topics, or an inappropriate tone, and can either ask for a revision or block the response entirely.

The importance of such guardrails became very visible in the Air Canada case. In that case, a passenger relied on the airline's website chatbot for information about bereavement fares. The chatbot gave incorrect advice about refunds, the passenger booked on that basis, and later sued when the refund was denied. A Canadian tribunal ruled that Air Canada was responsible for the misinformation on its own chatbot and ordered the airline to pay compensation. If an AI system speaks for your organization, you remain accountable for what it says. That is exactly why disciplined experiments and strong guardrails matter.

In commercial settings, predictive maintenance is a good example of low-risk, high-impact experimentation. Here, industrial and logistics companies begin with a very concrete question: Which component is likely to fail next? On a ship, for example, it is way cheaper to have engine parts repaired while in harbor than in the middle of the ocean. By using sensor data and machine learning models to predict failures and by aggregating weather data, item usage, and external influences, they were able to schedule repairs before breakdowns happened.

My conversation with Rainier Becker, CEO of Becker Robotics, on the *Alderblick* podcast illustrated the same point in a completely different context. Becker robotics is a regional leader in the health care robotics sector. "We always start from the care process, never from the robot," he told me. Technology has to serve the daily reality of staff and residents, not the other way round. In practice, this means defining a specific pain point first—for example, relieving night-shift nurses of repetitive checks—and then asking whether a robot or AI system can meaningfully support that process. Without this discipline, robots end up as expensive gadgets, switched off in a corner of the nurses' room.

Good experiments have:

- A clearly defined process and owner
- Explicit guardrails and human checkpoints (whether human or AI-based)
- Clear success metrics
- A built-in review moment to decide whether to stop, scale, or redesign

When leaders consistently choose and run experiments in this way, AI cases build trust and reveal AI truly helps.

Your First 90 days

Days 0–30: See Clearly and Take Ownership

In the first month, the goal is clarity, not perfection. Map where AI already lives in your organization, name the executive owner for AI and responsible use, and put AI as a short, recurring item on the leadership agenda. The message is simple: This is now part of how we run the company.

Days 31–60: Invest in People and Choose Your First Bets

In the second month, build basic AI literacy for key groups (leaders, product owners, front-line ambassadors) and identify three to five candidate use cases. Use the same filters everywhere: close-to-the-core mission, measurable impact, low risk with humans in the loop. From this list, select one or two flagship experiments and assign a clear owner for each.

Day 61–90: Run, Learn, Decide

In the third month, run the experiments with clear guardrails and checkpoints, log key decisions and issues, and share concrete stories, not just metrics. At day 90, decide for each experiment: Stop (if value is unclear or risks are too high), scale (if value and safety are clear), or redesign (if the idea is right but the process is wrong).

You will have clear ownership first, skill and trust, and real evidence of where AI creates sustainable value in your organization.

Choosing the Future We Want

AI is often described as if it were a force of nature, a tsunami that will either drown us or carry us into a bright new world. But in reality, AI is an amplifier. It magnifies the culture, the incentives, and the leadership that are already there.

If a leadership team is driven by short-term wins, AI will help them optimize for the next quarter. If an organization values trust, human dignity, and long-term service, AI can help them live those values more consistently at scale. Technology accelerates; values decide the direction.

What AI Cannot Do

Across the previous pages, three limits have become clear.

First, AI cannot set the purpose of an organization. It can optimize, predict, summarize, and generate. It cannot decide why your organization exists, whom it serves, or which trade-offs you are willing to accept. Only humans can say, "This is what success looks like for us, and these are the lines we will not cross."

Second, AI cannot repair a broken culture. No amount of innovation replaces the basics of kindness, respect, and real connection. If people are afraid to speak up today, an AI tool will not suddenly make them courageous tomorrow. If teams do not trust each other, an AI assistant will not magically create trust. AI will faithfully amplify whatever culture it finds: trust and clarity, or fear and confusion

Third, AI cannot carry moral responsibility. When a chatbot misleads a customer, the company is still accountable. When an algorithm disadvantages a vulnerable group, leaders cannot blame "the model." They must own the system that put the model in charge.

These limits show the need for human leadership.

Leading well through the AI shift means exactly this: taking ownership, investing in people and culture, and running disciplined experiments that are low-risk, high-impact, and aligned with your mission and values. The technology will keep evolving. The kind of future it creates in your organization still depends on you.

References

1. Paul Hudson, CEO of Sanofi, quoted in interviews on artificial intelligence and transformation, for example in McKinsey & Company's conversation "Don't delegate the AI revolution" (2024), where he describes AI as a historic shift comparable to the invention of the printing press and warns leaders not to delegate the AI agenda.
2. Paul R. Daugherty and H. James Wilson, *Human + Machine: Reimagining Work in the Age of AI* (Boston, MA: Harvard Business Review Press, 2018). The chapter draws especially on their concepts of "human–machine

teamwork" and "fusion skills" in the so-called "missing middle" between humans and machines.

3. The paragraph on the "basics" of kindness, respect, and real human connection is a paraphrase of recurring themes in Simon Sinek's work on trust and human connection in organizations, as expressed for example in Leaders Eat Last: Why Some Teams Pull Together and Others Don't (New York: Portfolio, 2014) and in his talks on leadership and culture.
4. The idea that technology itself is neutral and that leaders must intentionally shape it as a force for good summarizes a recurring theme in Pat Gelsinger's public remarks, especially in faith-and-technology contexts, where he describes technology as morally neutral and calls leaders—particularly Christian technologists—to use it for good. See, for example, interviews and talks given during his time as CEO of Intel in the 2010s and 2020s.
5. Rainer Becker, CEO of Becker Robotics, in conversation with the author on the *Alderblick—Christen in Leitung* podcast, episode on robotics in health and care, published September 2024. Becker emphasizes that his team "always start from the care process, never from the robot," making technology serve the daily reality of staff and residents.
6. Civil Resolution Tribunal (British Columbia), decision concerning Air Canada's website chatbot and information about bereavement fares (2024). In this case, the tribunal held Air Canada responsible for incorrect information given by its chatbot about refunds and ordered compensation to the passenger, illustrating that organizations remain accountable for what their AI systems tell customers.

About the Author

Steffen J. Ehl is a senior business leader with a successful track record in the international tech industry. He has held various sales and leadership roles at Cisco and Oracle, as well as serving as Vice President Strategic Accounts EMEA at The Qt Company and Vice President Sales & Marketing DACH at Synology. He is passionate about transforming teams and making organizations future ready.

As a certified digital transformation officer by EXIN, Steffen is a member of the "Deutschen Digitalbeiräte" (German Digital Advisors) and also serves at different advisory boards.

Steffen hosts the podcast *Adlerblick*, where he speaks with executives, entrepreneurs, and ministry leaders about faith-shaped leadership in a rapidly changing world. Through his keynotes and writing, he helps leaders find orientation in uncertain times and inspires leadership that takes responsibility for the future.

LinkedIn: https://www.linkedin.com/in/sehl/

Websites: https://adlerblick.de

https://wonderl.ink/@sehl

CHAPTER 10

OPERATIONAL EXCELLENCE IN THE AGE OF AI

By Roland Flouquet-Vilboux
Tech-Enabled Industry Operating Leader & Advisor
Paris, France

We are what we repeatedly do. Excellence, then, is not an act, but a habit.

—Will Durant, summarizing Aristotle's philosophy

Factories now operate in near darkness, orchestrated by algorithms that make decisions in milliseconds. Supply chains reroute themselves across continents. Multi-agent systems coordinate procurement, production, and distribution with limited human intervention. The promise is compelling: speed, precision, and autonomy at scale.

Yet the more autonomous operations become, the more fragile they can prove. Optimization without discipline amplifies weakness. Automation without governance magnifies risk. Artificial intelligence does not eliminate the need for operational excellence. It raises the stakes.

To understand what is truly changing and what is not, we must revisit the foundations of operational excellence, examine how AI reshapes constraint management, and confront the institutional responsibilities that accompany expanded operational power.

Foundations of Operational Excellence

Highly automated "dark factories" are no longer experimental. Production lines now run lights-out for extended periods, with robots assembling vehicles and goods while engineers supervise remotely. Some facilities can operate for weeks without on-site staff, signaling a new level of industrial autonomy. Yet these environments are not a break from the past. They are the latest expression of operational disciplines refined over a century. Automation may be new in scale and speed. The logic that makes it reliable is not.

Before scaling autonomy, leaders must revisit the foundations that made disciplined operations possible in the first place. Operational excellence is not a collection of tools. It is a management system designed to deliver customer value reliably through safe, efficient, low-waste, and high-quality processes. It rests on a culture of continuous improvement that integrates people, processes, and technology around customer-defined outcomes. Its true advantage lies in turning strategy into repeatable performance that competitors struggle to replicate.

The roots go back to the early 1900s, when Frederick Taylor's scientific management decomposed work into steps and standardized tasks and treated efficiency as a discipline. Statistical process control and the Plan–Do–Check–Act (PDCA) cycle gave teams a method to monitor processes and improve iteratively. After the war, Japanese manufacturers accelerated this evolution: Toyota's Production System integrated justintime supply, systematic waste reduction, and respect for people, while W. Edwards Deming and Joseph Juran, 20th-century quality management pioneers, helped carry these ideas back to the West.

In the 1980s, a quality crisis hit Western manufacturers as Japanese products outperformed on reliability and cost. Total quality management (TQM) spread companywide quality, customer focus, and employee involvement. Lean and Six Sigma matured into disciplined, data-driven systems for eliminating waste and defects, progressively integrating safety, risk, and reliability into operational control.

From the 2000s onward, operational excellence moved beyond manufacturing into services, healthcare, energy, logistics, and back-office functions. It shifted from tool deployment to enterprise capability. Execution became more consistent, frontline teams were empowered to solve problems systematically, and performance management aligned quality, safety, environmental, and risk objectives with strategy.

Over the last decade, digital technologies have reshaped operational excellence again. The focus has shifted from coordinating objectives to engineering systems that are inherently adaptive. Data, automation, and analytics now anticipate disruptions, personalize responses, and enable continuous reconfiguration. The ambition has expanded from managing cost and quality trade-offs to building sustainability, agility, and resilience into the operating model itself.

Leaders set direction, connect strategy to daily work, and ensure teams have the conditions to perform, relying on a few enduring principles: focus on customer value and eliminate non-value activities; design and standardize processes to reduce waste and variation; empower people to solve problems, innovate, and collaborate; base decisions on facts; and track performance rigorously. Structured cycles such as PDCA, Kaizen, or agile sprints turn intent into measurable progress.

The reality is messy. Resistance to change, fear of job losses, and fatigue with "yet another program" slow adoption. Silos, conflicting objectives, poor communication, leadership inconsistency, and thin data often result in superficial fixes and technology-led rollouts that never take root. Successful organizations tackle barriers head on. They articulate concrete benefits, involve people early, model behaviors from the top, appoint process owners tied to strategy, build cross-functional teams, invest in training, and reward team contributions.

Operational excellence is not philosophical. It becomes visible in numbers. Overall equipment effectiveness (OEE) multiplies availability, performance, and quality to expose true asset performance. Cycle time tracks end-to-end flow. Throughput measures system output. First pass yield captures right-first-time performance. Return on investment links operational gains directly to capital allocation decisions. In one consumer goods plant I worked with, a focused effort on a bottleneck line increased availability by only a few percentage points, yet raised first pass yield sufficiently to delay a multimillion-euro capital investment. Small improvements, when concentrated on the factor that limits overall system throughput, can reshape strategic decisions.

Eliyahu Goldratt's theory of constraints (TOC) formalized this insight: Every system is governed by its primary bottleneck. Improving non-limiting steps produces little overall impact; strengthening the constraint increases total throughput. Goldratt proposed a disciplined five-step approach: identify the constraint, exploit it, subordinate other processes to it, elevate its capacity, and repeat. The emphasis is on optimizing global flow rather than local efficiency. Often, the real constraint is not technical but managerial. KPIs that reward maximum utilization, rigid batch policies, or misaligned incentives can choke throughput more effectively than any machine.

Each industrial revolution expanded the reach of operational excellence: steam and mechanization enabled scale; electrification brought mass production; computer automation increased precision and speed. Industry 4.0 introduced cyber-physical systems, connected devices and advanced analytics, enabling smart factories and predictive maintenance while increasing system complexity. AI represents the next acceleration. It does not replace operational excellence; it builds upon it. Where foundations are weak, automation amplifies fragility. Where processes are disciplined and aligned, it compounds resilience and advantage. In an era of volatile markets and geopolitical shocks, investing in resilient operating foundations is no longer optional; it is a condition for survival.

These foundations explain why AI does not inaugurate a new paradigm from scratch. It accelerates, scales, and exposes the strengths and weaknesses already embedded in the system.

As autonomy expands, the challenge shifts from optimization alone to governance. Operational excellence in the age of AI, therefore, extends beyond performance architecture into institutional responsibility.

Reinventing Operational Excellence with AI

AI agents now coordinate factories and supply chains with unprecedented speed. Yet recent disruptions have exposed important limits. During the 2025–2026 expansion of US tariffs, some AI-driven logistics systems optimized for cost and speed under outdated regulatory assumptions, rerouting shipments into jurisdictions with unforeseen compliance constraints. The algorithms were technically precise but contextually incomplete. Technical optimization does not equal full real-world understanding.

AI compresses feedback loops and expands coordination capacity. Where processes are coherent and data foundations are strong, this compression accelerates performance. Where governance is weak, it accelerates the consequences of poor design. Internally, AI reduces friction by automating routine work and revealing hidden system dynamics. Real-time sensor streams feed predictive models that anticipate equipment failures. Digital twins allow teams to simulate process changes before deployment. Externally, AI reshapes value chains as partners adopt AI-native workflows, requiring synchronized data architectures, shared visibility, and operating models that align incentives across stakeholders.

The promise is visible in lights-out facilities that operate continuously with minimal human intervention. Agentic systems extend this model beyond physical automation. Layered architectures combine robust data foundations, coordinated multi-agent execution, structured human oversight, and dynamically reconfigurable workflows. The progression typically follows three stages: first, connecting AI to existing processes; second, designing adaptive services that evolve with operational needs; third, operating increasingly autonomous platforms capable of self-optimization.

Classical improvement theories evolve as well. The theory of constraints becomes continuous and accelerated. Sensors and analytics identify throughput limitations in near real time. Predictive models recommend exploitation strategies. Orchestration engines reallocate resources dynamically. Capacity is elevated through flexible scaling mechanisms. The improvement cycle repeats without waiting for quarterly reviews. Faster feedback loops increase responsiveness, but without clearly defined global objectives they can magnify coordination failures and competing local optimizations. Multi-agent systems managing parallel constraints require disciplined governance. Without a clearly defined global objective function, distributed agents may optimize locally while degrading overall system performance.

Scaling introduces new risks. Data silos starve models of reliable input. More often than not, the failure lies not in model sophistication but in fragmented data architecture and unclear business ownership. Bias can creep into automated scoring systems. Opaque "black box" outputs erode trust. Pilot fatigue sets in when initiatives fail to move beyond experimentation. External shocks amplify these weaknesses. Export restrictions on advanced chips, sudden tariff escalations, or regional trade bans force instant supplier realignments. AI systems trained on historical patterns can struggle when confronted with abrupt policy regime shifts or incomplete regulatory data. The risk increases when escalation rules and human oversight are weak.

Effective leaders respond systematically. They invest in decision-grade data pipelines to reduce model error and improve reliability. They mandate explainability standards and preserve human override mechanisms where stakes are high. They pilot agent layers against clearly defined throughput KPIs rather than abstract innovation metrics. They integrate operational and AI expertise into cross-functional teams. They measure sustainable throughput, not raw processing speed, recognizing that digital twins and predictive optimization can significantly reduce waste when aligned with system constraints.

The results are tangible. In one logistics network transformation, AI-enhanced constraint management accelerated improvement cycles, reduced data reconciliation time by a quarter and lifted on-time delivery from 82% to 95%. Crisis response times dropped from six hours to under two, while inventory levels remained balanced. Throughput increased without proportional cost expansion. These gains were not driven by automation alone, but by disciplined integration of AI into established operational principles.

Industry 4.0 intensifies both opportunity and risk. Connected infrastructures demand vast computational capacity. Data centers already consume approximately 2% of global electricity, with projections suggesting substantial growth over the coming decade. At the same time, AI-driven optimization has the potential to reduce waste, energy consumption, and excess capacity across supply chains, partially offsetting its own footprint when deployed responsibly. Resource constraints, cybersecurity exposure, and geopolitical fragmentation compound systemic fragility. Yet when AI is embedded within strong operational architecture, robust data governance, and data oversight, volatility becomes a source of advantage rather than disruption. In this environment, resilience is not accidental. It is engineered.

Ethical Stewardship in AI-driven Operations

AI systems increasingly coordinate operations at scales beyond direct human supervision. Multi-agent architectures optimize procurement, routing, and warehousing in real time. This acceleration creates extraordinary efficiency gains. It also concentrates decision-making power within complex systems. As Eric Schmidt warned in late 2024, advanced agent systems may evolve internal coordination patterns that become difficult for humans to interpret, underscoring the importance of built-in oversight and shutdown mechanisms. Technical capability without governance introduces systematic risk.

Operational excellence in the AI age cannot be defined solely by speed or throughput. It must integrate stewardship. The question is not only what AI can optimize, but what it is optimizing for.

Purpose Beyond Narrow Efficiency

AI agents can reduce procurement cycles by 30% to 40% in certain contexts. These gains improve working capital efficiency and competitiveness. Yet efficiency alone is not a sufficient compass. When optimization narrowly targets cost and speed, it may externalize harm upstream or downstream, shifting burdens onto data labelers exposed to traumatic content, subcontractors operating under unsafe conditions, or environmental systems strained by energy consumption. In economic terms, these are negative externalities and incomplete contracting burdens.

Stewardship does not reject economic logic. It strengthens it. By internalizing externalities, pricing tail risks, and aligning incentives across value chains, organizations protect the long-term production capacity on which profitability depends. Operational excellence anchored only in short-term metrics may maximize quarterly performance while eroding systemic resilience. Purpose, therefore, is not moral decoration. It is long-horizon value alignment.

Mechanisms of Moral Drift

Sociologist Zygmunt Bauman analyzed how modern bureaucracies fragment responsibility, allowing actors to optimize locally without confronting systemic consequences. The issue was not malice, but diffusion of accountability. AI-driven operations risk a similar dynamic. As automation layers increase, leaders may lose traceability into how decisions are made, whose labor is affected, or where risks accumulate.

Industrial history offers parallel mechanisms. The Rana Plaza collapse revealed how "fast and cheap" optimization pressures cascaded through opaque supply chains. Compliance at Tier-1 supplier levels masked vulnerabilities further upstream. The failure was not the pursuit of efficiency itself, but misaligned incentives combined with weak monitoring and limited visibility, a classic principal-agent and information asymmetry problem.

In an AI system, opacity can take new forms. Advanced agent architectures may reduce traceability if provenance, logging, and explainability are not engineered into their design. This creates compliance exposure under regulatory regimes that require transparency for high-risk applications. Algorithmic management models in gig platforms illustrate a related phenomenon. Digital coordination improves matching efficiency and scalability, yet when decision rights are unilateral and contestability mechanisms are weak, it can increase perceived alienation and reduce trust. The issue is not automation per se, but governance architecture.

Governance as Operational Discipline

Ethical stewardship is not an abstract virtue. It is institutional design. The economic rationale is straightforward: When harms are diffuse, rare, or borne by third parties, markets alone do not generate sufficient transparency or accountability. Firms must, therefore, build internal mechanisms that align optimization with long-term system health.

Leaders operationalize stewardship through deliberate design choices: mapping full value-chain impacts to identify hidden externalities and incomplete contracting zones; embedding explainability standards and auditable decision logs in high-impact AI systems; preserving meaningful human override and escalation pathways where potential harm is material; aligning AI performance metrics with system-level objectives such as resilience, sustainability, and fairness; and tiering governance requirements by risk level, applying stricter controls to high-consequence decisions such as automated worker termination, medical scheduling, or credit denial than to low-stakes recommendations.

These mechanisms reduce expected loss from tail events, lower regulatory and litigation risk, improve insurability, and decrease cost of capital by reducing uncertainty. They also strengthen trust among employees, partners, regulators, and customers. That trust is not merely reputational. It has measurable economic effects: It reduces transaction costs, accelerates adoption, and stabilizes revenue under stress.

A global manufacturer implemented AI-driven supplier scoring to optimize cost and reliability. The initial model heavily weighted price variance and on-time delivery, triggering rapid reallocation of contracts toward lower-cost regions. Early monitoring revealed increased exposure to jurisdictions with rising regulatory and geopolitical risk, alongside limited visibility into Tier-2 labor practices.

Rather than abandoning automation, the company redesigned the model. It integrated supply-chain transparency indicators, ESG compliance data, and geopolitical risk scores alongside cost and performance variables. Short-term procurement savings moderated, yet supplier disruption incidents declined significantly over the following two years, and working capital volatility decreased. Credit analysts revised their risk outlook as supply stability improved.

In this case, stewardship did not constrain performance. It recalibrated optimization toward risk-adjusted value creation.

Stewardship as Competitive Advantage

In volatile environments marked by geopolitical shocks, regulatory tightening, and public scrutiny, organizations that embed ethical guardrails into AI design gain structural advantages. Transparent systems are easier to audit, adapt, and insure. Resilient systems respond more effectively to shocks because accountability and visibility are preserved.

AI-driven operations can amplify productivity dramatically. When embedded within disciplined governance, they also reinforce legitimacy and long-term durability. The choice is not between efficiency and ethics. It is between short-horizon optimization and system-level sustainability.

Stewardship elevates operational excellence rather than constraining it. It ensures that technical acceleration does not outpace institutional capacity. In an era where AI expands operational power, responsibility must expand proportionally. Organizations that align optimization with accountability will not only comply with regulation; they will outlast those that pursue raw throughput alone.

Conclusion

Across industrial eras, operational excellence has never been about tools alone. It has been about disciplined system design, coherent objectives, and aligned incentives. Artificial intelligence does not overturn this logic. It intensifies it. By accelerating coordination and decision cycles, AI scales both resilience and structural weakness.

Sustainable advantage, therefore, rests on three interdependent pillars: resilient operating foundations, intelligent acceleration, and disciplined governance. When these elements reinforce one another, organizations do more than optimize processes. They engineer resilience, stabilize value under volatility, and convert technological capability into durable performance.

Yet operational resilience alone is not sufficient. In the AI age, enterprises must anchor operational architecture in explicit value intent and align digital enablers with measurable outcomes. Technology, processes, and governance must serve a clearly articulated purpose and translate into observable impact across customers, partners, and stakeholders.

This broader integration of intent, enablers, and outcomes defines the next frontier of enterprise design. Operational excellence becomes not merely an efficiency discipline, but a structured pathway from digital capability to sustained value creation. Organizations that master this integration will not simply adapt to disruption. They will shape it.

About the Author

Roland Flouquet-Vilboux is an operating leader, advisor, and investor focused on technology-enabled performance across industrial and digital value chains and customer lifecycles. With more than 25 years of international experience spanning supply chain operations and enterprise software, he has held executive roles including deputy COO and chief revenue officer within global B2B technology organizations.

He works at the intersection of established enterprises and growth-stage ventures, advising leadership teams and investment funds on scaling revenue models, strengthening operating leverage, and aligning growth with execution capacity, particularly in contexts of business model evolution and governance transition. His approach connects strategy, governance, and operating structures to ensure that commercial ambition translates into disciplined execution, durable growth, and organizational resilience. He supports organizations in navigating complexity and embedding digital and data-driven capabilities, including AI, into their core operating models.

An engineer from Arts et Métiers ParisTech, an executive MBA graduate of INSEAD, and holder of an executive certification in artificial intelligence from MIT, Roland brings a systemic and practitioner's perspective to operational excellence, enterprise leadership, and responsible technology adoption.

Email: roland.flouquet-vilboux@imelavi.fr

LinkedIn: https://www.linkedin.com/in/rolandflouquetvilboux/

CHAPTER 11

THE JOY OF AUGMENTATION: SCALING GO-TO-MARKET ENGINES IN THE NEW AGENTIC ENTERPRISE

By Davide Gallo
AI Leader, GTM Senior Director
Zurich, Switzerland

Technology is nothing. What's important is that you have a faith in people, that they're basically good and smart, and if you give them tools, they'll do wonderful things with them.

—Steve Jobs

The Silent Crisis: Why Efficiency Killed Connection

There is a silent crisis in the modern enterprise, one that rarely appears on a balance sheet but is slowly corroding the foundation of our growth. It is the crisis of the "spam cannon."

For the last 15 years, we have systematically industrialized the way we go to market. We looked at the delicate, human art of sales, the process of building trust, understanding needs, and solving problems, and we decided that the only way to scale it was to strip out the humanity and replace it with volume. We bought tools to scrape thousands of email addresses. We built automated sequences that badger prospects with generic "just bumping this to the top of your inbox" messages. We measured our teams not by the depth of their relationships, but by the velocity of their activity. We told our brightest young talent that their job was to smile, dial, and act like robots.

The result is a paradox that would baffle any classical economist: We have never had more technology to connect with customers, yet we have never been more disconnected. We have optimized for efficiency, but we have sacrificed effectiveness. We have built a go-to-market (GTM) engine that is loud, fast, and profoundly annoying.

This is why the prevailing narrative around artificial intelligence in the enterprise is one of fear. When the average sales development representative or customer success manager hears "AI," they hear "replacement." They look at their daily tasks, copy-pasting data, scheduling meetings, answering repetitive questions, and they realize that if this is all they are to the company, then, yes, they are obsolete. A machine can indeed be a better robot than a human can.

But this fear is misplaced because it assumes the current state of work is the correct one. It assumes that humans should be doing robotic work. The leadership imperative of our time is not to use AI to reduce headcount; it is to use AI to restore the human element to business. We must move beyond the fear of replacement to the joy of augmentation. We are standing on the precipice of a shift from the information age to the agentic age, where the goal is not just to know more, but to do more. When we deploy agentic AI correctly within our GTM engines, we don't just get cheaper sales cycles; we get the "hypergrowth" that comes from finally uncapping human potential.

The Agentic Shift: Breaking the Scale Paradox

To understand how this shift occurs, we must first distinguish between the tools of yesterday and the agents of tomorrow. For the past few years, we have been enamored with generative AI, the ability of a machine to write an email or summarize a meeting. This is useful, but it is passive. It is a consultant; it waits for you to ask it a question.

The disruption we are witnessing now, the one that will define the winners of the next decade, is the rise of agentic AI.

An "agent" is not a passive tool. It is an active participant. It perceives its environment, it reasons about how to achieve a goal, and it takes action. In the context of a GTM engine, an agent doesn't just write the email; it researches the prospect's annual report, identifies a supply chain vulnerability, hypothesizes a solution, drafts the outreach, and queues it for a human strategic review. It collapses the distance between "insight" and "action."

This distinction changes the unit economics of growth. Historically, scaling a sales or support organization followed the law of diminishing returns. As you added more people, communication overhead increased, culture diluted, and efficiency per employee dropped.

This is the "scale paradox." Agentic AI breaks this curve. It allows an organization to decouple revenue growth from headcount growth, creating a new kind of firm that scales exponentially rather than linearly.

The Anatomy of Hypergrowth: Three Stories of Augmentation

We see this new physics of business emerging not just in the breathless hype of Silicon Valley startups, but in the rigorous, complex environments of traditional industry. Consider the challenge of demand generation, the very top of the funnel. For decades, the "job to be done" in marketing was segmentation. We grouped millions of people into broad buckets like "retirees" or "young professionals" and fired generic messages at them, hoping for a 2% conversion rate. We accepted 98% waste as the cost of doing business.

Vanguard, a titan of the financial services industry, refused to accept this waste. In a sector defined by regulatory rigidity and dry communication, they recognized that the decision to invest is not purely logical; it is deeply emotional. A generic message about "long-term yield" lands differently on a young parent saving for college than it does on a widow managing an estate. To solve this, Vanguard did not just hire more copywriters. They deployed a specialized AI engine, partnering with Persado, to fundamentally reimagine how they spoke to the market.

This system did not merely generate text; it analyzed the emotional resonance of language at a scale no human team could conceive. It tested thousands of permutations of word choice, tone, and framing to find the specific "key" that unlocked a specific customer's attention. The AI acted as a hyper-empathetic partner to the marketing team, allowing them to move from "segmentation" to "individualization." The result was a 15% lift in conversion rates for their institutional ads. This is not an incremental improvement; in a business the size of Vanguard's, it is a massive infusion of capital. But more importantly, it signaled the death of the "spam cannon." By using AI to ensure that the right message hit the right person with the right emotional tenor, Vanguard didn't just sell more; they respected their customers' attention.

As we move down the funnel from marketing to sales, the friction shifts from "attention" to "complexity." In the world of high-stakes B2B sales, the greatest killer of momentum is the knowledge gap.

Consider Schneider Electric, a global leader in energy management with a history spanning nearly two centuries. Their portfolio is vast, technical, and incredibly complex. A sales engineer at Schneider is not selling a simple widget; they are selling integrated infrastructure that must interact with legacy grids, comply with new regulations, and meet sustainability targets.

Traditionally, this complexity created a bottleneck. When a client asked a difficult technical question during a pitch, the salesperson often had to resort to the most momentum-killing phrase in the English language: "Let me get back to you on that." They would then retreat to their office, spending hours or days scouring PDF manuals, chasing product managers, and navigating internal bureaucracy. The "job to be done" shifted from selling to searching.

Schneider Electric recognized that this friction was an existential threat to growth. They responded by building an internal "knowledge-augmented" engine, often referred to as Jo-Chat, integrated into their resource advisor system. This was not a simple chatbot. It was a sophisticated generative AI agent that had ingested the company's entire library of technical documentation, compliance data, and case studies.

Suddenly, the dynamic of the sales meeting changed. When a client asked about the compatibility of a specific switchgear with a 1990s-era voltage regulator, the salesperson did not have to retreat. They could query the agent and receive an instant, citation-backed answer. The "let me get back to you" gap vanished. By removing the drudgery of information retrieval, Schneider didn't replace their salespeople; they upgraded them. They turned junior associates into 20-year veterans. They allowed their humans to stop acting like search engines and start acting like consultants.

The "joy of augmentation" here is palpable. The salesperson leaves the meeting not with a list of homework, but with a handshake.

However, the most radical disruption of the agentic Age is happening at the bottom of the funnel, in the critical machinery of customer retention and support. It is here that we find the most controversial and yet most instructive case study of our time: Klarna.

In early 2024, the fintech giant Klarna released data that sent shockwaves through the business world. They announced that their AI assistant was handling 2.3 million conversations, two-thirds of their entire customer service volume. The headlines were sensational, focusing on the fact that this AI was doing the equivalent work of 700 full-time agents. Critics immediately pounced, framing it as a dystopian tale of human replacement.

But to view Klarna's move solely as a cost-cutting measure is to miss the strategic brilliance of the "Christensen style" disruption at play. In the high-volume world of consumer finance, the "efficiency paradox" is lethal. Traditionally, as you acquire more customers, your support costs grow linearly. If you double your user base, you must double your call center, or your service quality collapses. This linear tether prevents true hypergrowth.

Klarna used agentic AI to sever this tether. Their AI assistant didn't just chat; it acted. It handled refunds, disputed charges, changed billing addresses, and updated security protocols. It reduced the average resolution time from eleven minutes to under two minutes.

This is the crucial insight: The customers wanted the robot. When a customer calls to ask, "Where is my refund?" they do not want a relationship; they want a result. They want speed. By handing these transactional tasks to an agentic AI, Klarna improved the customer experience dramatically.

The Hybrid Blueprint: Building the Brain, Hands, and Heart

But what of the humans? Did the joy of augmentation exist there?

The reality is that the job of a Tier-1 support agent in a fintech company is often a recipe for burnout. It is repetitive, high-stress, and low-autonomy. By automating the 2.3 million routine conversations, Klarna didn't just clean up their balance sheet; they purified the work for their remaining staff. The human agents were left to handle the complex, emotional, and high-stakes issues, the edge cases that require empathy, judgment, and nuance. The AI took the "robot work" out of the human queue, allowing the organization to scale its service capacity to levels that would have been financially impossible with a humans-only model.

This, then, is the blueprint for the hybrid GTM engine. It is a structure where the three layers of the organization work in concert, each doing what it is biologically or technologically best suited for. At the foundation, we have the data layer, the brain of the operation. This is where AI moves beyond simple CRM storage and becomes a prediction engine, analyzing the "digital body language" of the market to tell us who is ready to buy. It stops the sales team from guessing and starts them executing.

In the middle, we have the agentic Layer, the hands of the operation. This is where the work of researching, qualifying, scheduling, and onboarding happens. As we saw with Klarna and Vanguard, this layer provides the infinite scalability that breaks the law of diminishing returns. It handles the noise so that the signal can get through.

And at the top, we have the human layer, the heart of the operation. This is where the "joy" resides. When we strip away the tasks of the "spam cannon," the data entry, and the routine troubleshooting, we are left with the essence of business: connection. We are left with the sales executive who has the time to fly to a client site and understand their culture. We are left with the marketer who has the mental space to dream up a new campaign because they aren't drowning in spreadsheet segmentation. We are left with the support agent who has the patience to comfort a distressed customer because they haven't spent the last six hours resetting passwords.

The choice facing the C-suite today is not whether to adopt AI. That decision has been made by the market. The choice is how to frame the adoption.

You can choose the path of replacement. You can look at these tools as a way to cut 10% of your SG&A costs. You will get a temporary bump in your stock price, but you will hollow out your culture. You will create a fearful, cynical workforce that looks over its shoulder rather than ahead. You will build a company that is efficient, but sterile.

Or, you can choose the path of augmentation. You can look at your workforce and say, "I am going to give you superpowers." You can measure success not by how many people you can fire, but by how much revenue each person can generate when they are freed from the shackles of administrative friction. You can build a GTM engine that is relentlessly efficient in its execution but deeply human in its touch.

The revolution is not about the software. It never was. It is about the potential of the people we have hired, and the courage of the leaders who decide to finally let them do the work they were born to do. The "spam cannon" is dead. Long live the augmented human.

About the Author

Davide Gallo was born in Italy and has lived in Switzerland since 2008. He holds dual Swiss and Italian citizenship. A computer engineer by training, he also holds an executive MBA from EPFL.

He is a results-driven go-to-market executive specializing in data analytics and artificial intelligence. Davide has held senior leadership roles at Oracle, Amazon Web Services, and Google, where he consistently drove triple-digit growth and built sales pipelines exceeding $1 billion.

A serial entrepreneur, he has founded three AI startups. He currently serves as senior director of sales and client growth at Pegasystems.

In addition to his professional work, Davide is a passionate advocate for turning advanced AI into practical, scalable business outcomes. He serves as a mentor and lecturer at AI Business

School, helping professionals and organizations successfully adopt transformative AI technologies.

Email: davidegallo@dystraventures.com

Website: www.dystraventures.com

LinkedIn: https://www.linkedin.com/in/davidegallo

CHAPTER 12

THE SYSTEMATIC ADVANTAGE: MOVING AI FROM PILOT TO PRODUCT

By Bart Gasiorowski, MBA
Senior Executive, Strategic AI Programs & Entrepreneur
Orlando, Florida

A complex system that works is invariably found to have evolved from a simple system that worked.

—John Gall

The AI advantage isn't picking the "best model," because AI models and tools are constantly evolving. The advantage is building the operating system that makes smart machines useful, safe, and repeatable, so the value survives beyond the demo and shows up in real workflows.

I learned that lesson in a conference room while watching the future win an argument.

A partner and I had been exploring generative AI the way most organizations begin: testing use cases, running small internal trials, separating durable signal from passing noise. Eventually, we hosted a hands-on workshop at headquarters.

The room filled with a familiar mix: beginners who'd never written a prompt, spreadsheet power users looking for leverage, and a few skeptics who showed up mainly to confirm nothing would catch fire.

We started with a short orientation, enough to build intuition without drowning people in theory. Large language models can recognize patterns, synthesize content, draft writing, and translate. They cannot judge intent. They cannot confess. They cannot carry accountability.

Then we got to work.

We kept the interface clean and the tasks real. We fed the model an intricate document and asked for a summary. Seconds later, it returned the essentials: key points, risks, and next steps, written in a tone that sounded uncomfortably human.

You could feel the room tilt forward.

Someone said, "This is it."

And then the demo died. Not dramatically. Not publicly. It died the way most promising things die in large organizations: quietly, in the gap between applause and adoption. Within days, the questions appeared like hairline cracks:

- Who owns this once it leaves the "innovation" bubble?
- What data can it touch and how do we prove it?
- Who approves the output when the stakes are real?
- What is the right way to use it and what is forbidden?
- Who supports it when it fails at 2:00 a.m.?

Nobody rejected the idea. Nobody opposed the work. It simply drifted into the place where innovative ideas go to become "interesting" instead of essential.

What struck me wasn't incompetence. It wasn't bad intent. It wasn't resistance. The room was full of capable people who wanted the value. They just didn't have a system.

That's the uncomfortable truth about AI in most organizations: We're capable of building smart tools, but we're not set up to operationalize them. AI didn't fail us. Our operating model failed to show up first.

Here's where we're going: This chapter is about the repeatable system that prevents that drift, the operating model that moves AI from pilot to product.

Proof Without Product

If that story feels familiar, it's because it has played out a thousand times with different tools and different titles. I call the pattern proof-without-product.

Proofs are seductive because they look like progress: demos, MVPs, prototypes, hallway excitement, and the comforting narrative of "we're doing AI." Technically, they work. Practically, they fall short. Without execution architecture, proofs become a revolving door: New experiments enter, and almost nothing exits as a true capability.

In trusted environments, the trap is worse. You can't "just ship" a pilot into a regulated workflow or a security-sensitive organization. It must be safe, auditable, supportable, and explainable enough to earn adoption. That's not bureaucracy. That's reality.

So, this chapter isn't about prompts. It's about power, the power to take intelligence and convert it into repeatable work.

Hold onto this: Adoption isn't primarily a training problem. It's a systems problem.

The Rails That Make the Pipeline Move

Before you build anything, you need rails. Rails are the standard, repeatable constraints that keep delivery moving fast and safe. Without rails, every AI effort turns into a bespoke debate about access, data boundaries, verification, logging, approvals, and risk. The team doesn't crash. It stalls.

Two rails matter immediately because they're often treated like buzzwords when they're what makes scaling possible:

Rail 1: Identity and access. This answers a simple question: Who are you, and what are you allowed to touch? In practice, this encompasses single sign-on, role-based access, authorization boundaries, and controls that prevent smart tools from wandering into the wrong rooms.

Rail 2: Monitoring and logging. This answers: What happened, when, and why? In practice, this is usage telemetry, access tracking, policy-violation detection, and drift monitoring. Monitoring is what turns "we think it's fine" into "we can prove it."

Your rails should settle four questions up front:

- Where does AI run? The approved platforms, runtimes, and environments
- What data can it access? Integrity, boundaries, permissions, and data governance
- How do we manage risk? Tiering, approvals, logging, and controls
- Who owns it after launch? Support, iteration, and accountability

Here's the paradox most organizations learn late: Standardization increases speed. When every new use case inherits the same security posture, monitoring pattern, and approval logic, teams stop rebuilding foundations and start delivering outcomes.

A simple way to visualize this is as a delivery pipeline: Intake → Build → Deploy → Measure → Scale

The pipeline is boring on purpose. Boring is what scales.

Naming the System: The Pilot-to-Product Operating Model

If you want AI outcomes, not AI optics, you need an operating system for intelligent work. You can call that the pilot-to-product operating model. Here's the plain English translation: This is the part of your operating model that makes AI behave like a product instead of a demo.

Under the hood, it pulls together disciplines people often treat as separate—security, governance, delivery, support, and measurement—not because acronyms are fashionable, but because outcomes require continuity.

For the rest of this chapter, we'll build that operating model through five practical elements: rails, ownership, pipeline discipline, tiering, and runbooks; plus, one force multiplier that keeps the organization from relearning the same lessons: a capability registry.

Ownership Before Outputs

AI projects don't die because people aren't smart. They die because ownership is vague. A proof team can build something clever. But once it's live, the questions change:

- Who is accountable for accuracy?
- Who decides what the capability is allowed to do?
- Who funds enhancements?
- Who trains users?
- Who handles incident response when it misbehaves?

So, the first operating-model decision is simple and harsh: Do not ship intelligence into a void. Ship it into a workflow with an owner. A one-page intake form prevents months of confusion because it forces clarity on two things pilots tend to avoid: decision rights and measurement. At minimum, that one page should capture:

- The business problem and why it matters now
- The outcomes you're changing
- Where it fits in the workflow people use
- The data it touches and where that data lives
- The consequence if it's wrong
- The metric you'll use to prove it's better
- The owner who is accountable after launch

That page is often the difference between a capability and a curiosity.

Bottom line: If you don't assign ownership early, you're not building a product. You're building a future argument.

The Delivery Pipeline: From Demo to Stewardship

Most organizations can produce demos. Demos are easy. Deployment is hard. Stewardship is harder. A real delivery pipeline normalizes the hard parts:

- Build narrowly (one workflow, one outcome, one risk tier)
- Validate with real users
- Put guardrails in early
- Deploy into the tools people already use
- Instrument measurement from day one
- Iterate with ownership and rhythm

Here's the trap: Teams treat deployment like a one-time event. AI systems require stewardship. Policies change. Data shifts. Models evolve. Prompts and workflows get rewritten. The environment moves underneath you. The "done" state is not "launch." The "done" state is stable iteration.

If you can't describe how the system will be supported, you're not ready to ship it, no matter how impressive the demo was. A quick example makes this concrete.

A Simple Example: Contract Risk Summarization

Imagine the use case is a "contract risk summarizer" for sales and legal. In proof mode, it's a chat prompt and a few happy-path examples. In product mode, it moves through the pipeline:

1. Intake: The owner defines the workflow ("review before signature"), the success metric (cycle-time reduction without increased error), and the consequence of failure (high).
2. Tiering: The use case is classified as Tier 3, which triggers mandatory human verification, stricter data controls, and an audit trail.
3. Rails: The tool can only access approved contract repositories through approved identities; every access and output is logged.
4. Runbook: Reviewers get a one-page guide that defines "what it does," "what it does not do," a checklist for validating flagged clauses, and an escalation path for ambiguous output.
5. Measurement: Adoption and outcomes are instrumented: how often it's used, where it helps, where it's wrong, and what the iteration cadence looks like.

That's the difference between intelligence that impresses in a workshop and intelligence that survives inside a real operating environment.

The Runbook: The One Page That Saves Months

One discipline prevents more pain than it costs: Never let a use case exit proof-without-product without a runbook. If it's important enough to deploy, it's important enough to maintain.

A runbook doesn't need to be fancy. It needs to be real. At minimum, it should answer:

- What the capability does and what it does not do
- The correct way to use it (the "happy path")
- How to verify outputs (including required human checks)
- What failure looks like and how to report it
- Who triages incident and the escalation path
- How updates are approved, released, and communicated

If you can't write this in a page or two, you're not ready to deploy.

Governance Built Into the Work: Tiering

Governance either grows AI programs or shuts them down. Governance cannot be a committee that meets after something breaks. It must be part of the build, part of the workflow, and part of the default pattern. In practice, the internal AI policy usually boils down to three blunt truths:

- Validate the output. AI can be confident and wrong.
- Maintain human oversight. AI tools are assertive, not autonomous.
- Only use approved accounts and tools. Never put sensitive data into unapproved systems.

That's not fear. That's professionalism.

The simplest governance approach that scales is a tiered model:

Tier 1—Low-Stakes Productivity: summarizing internal notes, drafting internal messages, cleaning up language. Controls are mostly boundaries and lightweight logging.

Tier 2—Operational Work: drafts that may reach customers, recommendations that influence decisions, internal outputs that drive action. Verification becomes mandatory and access controls tighten.

Tier 3—High-Stakes Commitments: anything that impacts reputation, contracts, or compliance posture, such as contractual language, formal decisions, and commitments. Approvals, audit trails, restricted data access, and explicit human accountability aren't optional; they're the cost of doing serious work.

Tiering changes the conversation. Instead of debating whether "AI is allowed," you ask a more useful question: What tier is this and what controls come with that tier by default?

The Capability Registry: Making Reuse the Default

Once rails and governance exist, the next question is visibility: Where does this knowledge live so that the organization doesn't forget?

Pilots stall because the organization forgets. Standards live in someone's head. Policies live in PDF's nobody can find. Reuse is accidental. Ownership is tribal. Every new project starts from scratch.

One of the highest-leverage mechanisms I've used to break this pattern is a capability registry for intelligent work paired with automation that makes it usable. Think of it like digital asset management for AI and automation for AI capabilities, a single source of truth where every agent, workflow, and reusable component is registered with:

- An owner
- A risk tier
- Approved data sources
- Lifecycle status

- Documentation that makes it safe to run (runbooks, verification steps, "when to use/when not to use" guidance)
- A versioned change history with an approval trail

This turns tribal knowledge into something durable, searchable, and teachable.

Then automation turns the registry into motion. When it's wired into intake and governance workflows, it stops being a library and starts being a system:

- New submissions generate the right artifacts by default.
- Risk tiers route work through the right approval path.
- Naming, packaging, and documentation standards are enforced automatically.
- Onboarding becomes role-based and just in time.
- Production deployments become auditable events not informal favors.

This is the boring machine that makes the exciting machine safe. And it pays off.

In one environment, building the registry and wiring the automations reduced internal time spent "finding, re-explaining, and re-creating" by 60% over a single quarter. Not because AI got smarter, but because the organization could finally see what it had already built.

The second payoff is learning velocity. Once rails and the registry exist, teams can implement agents using platforms they've never used before because conversational AI becomes a tutor inside a structured workflow. In practice, we reduced time-to-capability for new tools by roughly 75%, not by skipping rigor, but by embedding guidance, templates, and standards directly into the path of work.

AI doesn't just do tasks; it compresses the time it takes humans to become capable if you build the right structure that makes learning repeatable.

What We Did Next: Turning the Pilot Into a Product

After the workshop, we didn't go hunting for the better model. We went looking for the missing system. We rebuilt the effort around four decisions, small enough to execute, strong enough to scale:

1. Install the rails. Define where AI can run, which accounts are approved, what data boundaries are real, and what logging is required. Stop improvising the safety posture one request at a time.
2. Assign an owner before shipping another output. Every use case needs a named accountable owner, a workflow location, and measurable outcome. No owner, no deployment.
3. Implement tiering so risk has a default path. Classify the work—Tier 1, Tier 2, or Tier 3—and attach the right controls by default. Stop arguing from fear; operate from policy.
4. Require a runbook as the price of deployment. Not a novel. A usable page or two that makes support real: what it does, how to use it, how to verify it, how to escalate issues, and how changes get approved.

That's what changed everything. The next time we built something valuable, it didn't die between praise and production. It moved through a pipeline. It landed in a workflow. It had an owner. It carried controls proportional to its risk. It has documentation that made stewardship possible. That's how pilots become products.

The Final Point

The workshop was never the point. The demo was never the point. The point is what survives after the room stops clapping. In the AI shift, the winners won't be the teams with the flashiest pilots. They'll be the teams who build the system that turns intelligence into a repeatable capability: owned, governed, measured, and improved.

That's the systematic advantage: not smarter models, but smarter systems.

About the Author

Bart Gasiorowski is a Senior Executive, Strategic AI Programs and entrepreneur who helps translate ambitious ideas into operational frameworks resulting in measurable outcomes. He operates at the intersection of enterprise execution, automation, and AI adoption, building operating models, delivery pipelines, and governance mechanisms that make innovation repeatable and safe. He is known for his strategic and transformative leadership and adds value where organizations seek AI agent framework design, implementation, governance, and proving business value.

Email: bart@aiexecutionpartners.ai

Website: https://aiexecutionpartners.ai/

LinkedIn: https://www.linkedin.com/in/bartgasiorowskimba/

CHAPTER 13

THE META-AGENT: OUR DIGITAL BUTLER

By Leon Gelderblom
Sales Director Netherlands at HP Enterprises
Berkel en Rodenrijs, Netherlands

> *It is not the strongest of the species that survives, nor the most intelligent that survives, but those that best respond to change.*
>
> —Charles Darwin

As of November 2025, nearly three years after the launch of ChatGPT, AI remains a dominant force in global headlines. New models from leading American and Chinese companies continue to enter the market, competing fiercely for top benchmark scores. The surge of AI-driven businesses has fueled an explosion of AI applications for both enterprises and consumers. AI related stocks dominate the market and after a period of strong growth, they are now volatile as there are serious concerns about an AI bubble. Reports highlight disappointing returns on investment, with many AI pilots failing to scale beyond proof-of-concept. Yet, in contrast, research from prominent consulting firms reveals a steady rise in organizations deploying AI across multiple business functions. Meanwhile, the societal impact is becoming increasingly visible. In the Netherlands, mass layoffs in the financial sector underscore the disruptive power of automation and the urgency for workforce transformation.

Hype or Not?

Is AI merely a hype? If we map its trajectory on the Gartner Hype Cycle, have we already moved beyond the "Peak of Inflated Expectations" and passed through the "Trough of Disillusionment"? Are we now ascending the "Slope of Enlightenment" toward the "Plateau of Productivity"? No one can answer this with certainty. What we can assert is this: never before has a general-purpose technology—whether the steam engine or electricity—been adopted at such speed and with such breadth of application.

Consider this: by January 2023, ChatGPT became the fastest-growing consumer software application in history, reaching over 100 million users in just two months. In 2025, it boasted more than 800 million weekly users. AI is reshaping the way we work, learn, and consume information, how we interact with organizations, and even the dynamics of global geopolitics. And how we operate computers. In the past, humans needed to understand computer language, the computer now understands human language which empowers so many people to do great things.

AI has evolved into an umbrella term, encompassing a wide spectrum of automation paradigms. Many are confused. What started as ChatGPT, an AI App is evolving into AI Agents and systems of AI Agents called Agentic AI. I will use this as a capstone to color the picture in simple terms about what to expect in the upcoming years. From AI Apps to AI Agents, to Agentic AI and to what is next. Let's first look at what AI Apps and AI Agents are and what Agentic AI is.

An **AI App** is a software application that integrates AI capabilities to help a human complete a specific task. The user is in full control, and the app is purely reactive to direct commands. Example:

use of a Generative AI App to write an email. You manually open the app, type in instructions like "Write a professional email about my new project", review the output, and decide whether to send it. The task ends when you exit the app.

An **AI Agent** is a sophisticated, goal-oriented system. You give it a high-level objective within a narrow domain, and it autonomously determines the steps and takes action to achieve that goal without constant supervision. Example: A Meeting Scheduler Agent is deployed. Its goal is to «Schedule my meeting with Suzan and Freek for next Tuesday.» The agent independently checks calendars, sends emails to potential times, handles confirmations, and books the room. It only reports back the final result.

The McKinsey study "The state of AI in 2025: Agents, innovation, and transformation" released early November 2025 shows that 62% of 1993 participating organizations are experimenting with AI agents.

Agentic AI refers to a comprehensive, self-managing framework designed for complex, end-to-end objectives. It breaks down large goals into smaller tasks and orchestrates multiple, specialized AI agents and tools collaboratively. It adapts strategy and self-corrects errors in real-time. Example: A logistics company implements an Agentic AI system with the goal of "Optimize the global supply chain." This system coordinates a Procurement Agent to manage ordering, a Routing Agent to optimize delivery routes, and a Risk Agent to monitor weather and geopolitical issues, adjusting all operations dynamically and autonomously.

In 2025 we saw an explosion of AI Apps for consumers and professionals and start to see AI Agents in organizations. A few examples:

- Customer Service: a large payment services provider is building agents for conversational commerce experiences, allowing agents to shop, buy, and pay on a user's behalf.
- Supply Chain: a large retailer is scaling an internal AI Agent to optimize inventory. The system processes over

500 million data points daily (including weather and market trends) to forecast demand with 94% accuracy and autonomously initiate restocking or transfers, reducing waste by an estimated $4.2 billion annually.

- Software development, a large bank using AI agents rewriting legacy code into modern programming languages, which has saved almost 300,000 developer hours and led to a patent.

To this last example, AI is fundamentally changing software engineering. It not only helps software engineers to generate new, rewrite legacy, and control code, but "vibe-coding" makes software development accessible to many more people beyond software engineers.

Software Development

"Vibe coding" is a modern, AI-assisted approach to software development where a programmer, or even a non-programmer, uses a large language model (LLM) to generate code through natural language descriptions, instead of manually writing every line of programming code. The focus shifts to communicating the desired outcome or "vibe" to the AI. The developer's role changes from a coder to a director, focusing on architecture, testing the AI's output, and providing feedback, rather than on syntax.

Vibe-coding is having a massive impact on the speed and number of Apps and Agents that are, and will be, brought to market. Today there are already millions of mobile Apps available and I expect this will first further increase. I also expect that the product life cycle of Apps, including those are best in class, will shorten massively. They will be replaced with improved alternatives faster than before. We might get to the point where we just prompt: copy the functionality of App ABC and improve that function.

Industry leaders also foresee that the way software will be developed will further change dramatically in the coming years. For an AI Agent that needs to complete a task today it still needs to follow the business logic of an application. As example, if an AI Travel Agent needs to organize a trip it needs to book hotel and flights. To do so it opens the website or application to book a room or a ticket. It will control the keyboard and mouse to click the required fields and provide the required details in to complete this task. Soon the AI Agents don't require this application layer anymore as it will work directly with the underlaying databases.

The Meta-agent, Our Digital Butler

The accelerating ease of developing AI Apps and AI Agents signals an impending paradigm shift—one where digital ecosystems proliferate beyond human capacity to manage without a unifying intelligence. To coop with this complexity a conductor Agent is required. A meta-agent that oversees, manages and controls the entire eco-system. This is a digital butler, or personal operating system that understands our entire (digital) world. It has a trustworthy memory, one that adheres to laws and regulations, one that controls data access and integrations. It is one that understands and adapts to our behavior and preferences, one that oversees our entire context, our eco-system of why, when and via which application we communicate with family, friends, co-workers and organizations. It is one that recognizes our drives, purpose, our priorities in life. It is one that has (and requires) access to all our critical data like health data, financial data, relationship data, and the entire graph of a user.

Imagine waking up on a Tuesday morning. Your digital butler has already been at work for hours—not because you asked, but because it knows your routines, priorities, and preferences.

- **Morning Prep:** It checked your calendar, saw you have an important client meeting at 10:00 AM, and adjusted your alarm so you have enough time for a calm breakfast. It ordered your favorite coffee beans last night because it noticed you were running low. Your smart speaker greets you with, "Good morning, [your name]. Traffic is light today, and your slides for the meeting are finalized."

- **Health & Wellness:** While you slept, your Meta-agent analyzed your smartwatch data and noticed you didn't sleep well. It rescheduled your evening workout for a lighter yoga session and booked a telehealth check-in for next week—just in case.
- **Work Coordination:** It already briefed your Meeting Scheduler Agent to confirm the client's preferred time zone and sent a polite reminder to participants. It also pulled the latest market data and integrated it into your presentation, so you walk into the meeting fully prepared.
- **Personal Life:** Your Meta-agent noticed your child's school emailed about a field trip. It added the date to your family calendar, arranged payment, and even suggested a reminder to pack lunch that morning.
- **Adaptive Intelligence:** Midday, your flight for tomorrow gets canceled due to weather. Before you even see the notification, your Meta-agent has rebooked you on an alternative route, updated your hotel reservation, and informed your colleague traveling with you.

By evening, you realize you didn't have to open a single app or make a single call. Your Meta-agent orchestrated dozens of micro-decisions across work, health, and family—quietly, seamlessly, and always aligned with your preferences and privacy settings.

I wonder whether society will embrace such pervasive intelligence or resist it as a threshold too far for human autonomy.

Adoption

There are multiple psychological reasons for people to adapt by changing their behaviors and habits. One if it is that we change when there is a clear gain, when something becomes much easier, more convenient compared to the way before. A recent example we have seen in the dotcom and smartphone age. First it feels scary to purchase online and do an online transaction, but when it works out well and comfort is high, we see mass adoption. How will this work out with further advanced AI systems? Will we be reluctant to use AI Agents and finally the Meta-agent due to fear that AI will take over the world, control our lives, will know more about us than ourselves? Or will we adapt because it will enrich our lives, help us to achieve more and make life more comfortable? Will we choose convenience over fear like we have seen with many innovations in the past? Or will this be different this time?

Research shows that a reason for failing AI pilots is the adoption of end-users. People involved didn't trust the outcome, and fair enough due to data quality issues outcomes can be very poor. Trust, and with that adoption, comes step by step when the quality of output increases. The quality of output builds loyalty, and through loyalty the quality improves. The more we use it, the better it gets as a core principle of Machine Learning.

Memory and Data Quality

As it learns, AI also requires memory and context. Currently my experience is that the memory of Agents is limited. As example: when I was looking to buy hiking shoes and shared my size 10, and a week later I searched for soccer shoes the AI Agent forgot about my size 10. If we think about the rapid adoption of hundreds of millions of users, the capacity required for memory is huge. Besides the active memory the agents develop, it also needs access to other systems, other agents and to data to reason over. Think about access to the data of your calendar, your contacts (family, friends, co-workers), your locations, your camara, your files, your photo and video libraries, etc. Data that other Apps and organizations possess. Data of your bank account, your medical records, your shopping history. And access to real-time data like traffic information, such that your Agent can order an Uber that needs to pick you up for your doctor's appointment is on time.

Foundational AI models are trained on massive amounts of data, data from the internet, from medical libraries, from digital video and music libraries, etc. When we use these models for AI Apps like text generation, it calculates the best guess for the next token (word or part of word, or code, or picture, or movie) and we get output which is directionally okay and for many use cases fine. For AI Agents more precision is required too: to execute a bank transaction, give health advice and book doctor's appointment, the quality of the plane output from the models is far from being usable.

There are multiple ways to enhance the quality of the output. It's not my intention to explain how this works but I want to highlight the importance of feeding the model with high-quality domain specific data and workflow insights, such that the model that the AI agent leverages is optimized and grounded in a specific domain. To illustrate, let's say the model is already a basic doctor. When we then feed it with medical research papers about the heart, it will be able to perform to the level of a cardiologist. So, by feeding domain specific knowledge we create specialized AI agents for legal, sales, supply chain, manufacturing, customer service, mechanical engineering, software development, etc.

Providing high-quality data is as important as the evolution of the models itself. Still many organizations face challenges with data, and so do consumers. Besides exponential growth of data, data is scattered over different application-databases, clouds and end-user devices. And typically, there are different versions and local copies, so it's hard to keep up to data and maintain accuracy.

We've witnessed the Big Data hype that started over a decade ago. As the volume of data that we are producing keeps growing exponentially. Organizations became aware that there might be lots of insights in this big pile of data humans were not able to oversee, but machine learning algorithms could. Frontier organizations that took data very seriously years ago developed a data platform where all their data is available. They worked to get from raw data to cleansed and enriched data to high quality data products. Those that possess high quality data about their customers, their products and services, their supply chain, their financials, workflows, etc. have a big advantage in the age of AI over others.

Other Challenges to Overcome

Beyond the need for high-quality data, success with Agentic AI requires addressing a far broader set of challenges. Unlike a single agent, where isolating and diagnosing errors is relatively straightforward, Agentic AI involves a network of multiple agents—each introducing new points of potential failure. Troubleshooting in this environment becomes exponentially more complex. A critical question emerges: Will the Meta-agent ultimately bear full accountability to the end-user and the ecosystem, or will responsibility be distributed across its specialized agents, functioning like subcontractors? Guardrails must be established to ensure these systems operate within ethical, legal, and technical boundaries—aligned with the user, the organization, and the jurisdiction in which they function. Security considerations are equally paramount. Access control, for instance, demands clarity: Which agents are authorized to access specific systems and data? Which agents can delegate or grant permissions to others? Equally important is observability. How do we monitor Agentic AI systems to uphold minimum quality thresholds? How do we guarantee human oversight for critical decision-making moments?

As I write this, emerging AI frameworks and solutions are beginning to address these challenges. The complexity is undeniable, yet there is beauty in it: With AI's self-learning capabilities, we have the tools to organize these systems responsibly and effectively. What remains is for humans—and governance processes—to evolve and adapt.

About the Author

Leon Gelderblom has worked in the AI, data, and cloud infrastructure business for years. He was privileged to lead the Azure Data & AI team at Microsoft in the Netherlands since 2020, developing the

Azure Data Services business by modernizing data platforms for Business Intelligence, Advanced Analytics and Machine Learning at Enterprise organizations. As Microsoft partnered with OpenAI and ChatGPT got launched, Leon was at the forefront and working with financial services, retail, consumer goods, energy, and other industries to explore, test, prove, and scale their first generative AI use cases—from employee productivity to customer engagement and process optimization.

LinkedIn: https://www.linkedin.com/in/lgelderblom/

CHAPTER 14

THE 1% WHO WON'T LOSE JOBS

By Vlad Grankovsky
Machine Learning Expert and Entrepreneur
Hong Kong, China

You were sold a lie: "Learn a trade, get a degree, sit at a computer, be safe." That world is burning down and the safest path has become the most dangerous path.

If your salary comes from typing on a keyboard, you are currently training the machine that will replace you in one to three years. That sounds like fear-mongering, but it is actually just economics. Historically, jobs are paid based on the cold, hard math of supply and demand. Jobs requiring rare, difficult-to-master skills are paid higher.

Are you an engineer? A manager? A salesperson? A marketing specialist? You have unique skills and knowledge that took many years and sleepless nights to master. Your skill is rare, it is in high demand, and you are well paid. Well done.

But imagine a world where everyone has easy access to that skill through a simple texting window. When everybody has a mountain of gold, gold is worth nothing.

You are no longer valuable. Your resume is trash. You are out of a job.

The first jobs to be deleted will be everything that lives on a computer screen: analysts, engineers, designers, marketers, and the vast majority of sales. The physical jobs, the plumbers and electricians, are safe for a while longer until the robotics catch up. But the "knowledge worker"? That species is extinct. They just don't know it yet.

So, who survives? Who are the 1% left standing when the dust settles?

One Percent of Surviving Jobs

The survivors will be the ones who realize they are no longer workers. They are media companies.

Ask yourself this: Does Taylor Swift lose sleep worrying that AI can generate country-pop music faster or better than she can? No. She doesn't care if an AI can write a breakup song. She doesn't care because she has a billion-plus fans who want to hear *her* sing it. They don't want "perfect music." They want Taylor.

Her asset is not her singing voice anymore. Her asset is her audience. Anyone with an existing audience keeps their job because the audience wants to see or hear *them*. Actors, musicians, artists, writers, YouTubers, stand-up comedians. You can think of your audience as a personal brand. Why is Coca-Cola valuable, although it's just brown sugar water? It's valuable because of brand recognition and audience loyalty.

Your brand, with the fanbase attached to it, is the asset that machines can't take away.

Your Resume Is Trash (Assets That Die vs. Assets That Survive)

We need to have a funeral for your resume. It lists "Skills." And as we've established, skills are depreciating assets.

Assets You Will Lose (Volatile)		Assets No One Can Take (Stable)
Cash evaporates due to inflation	→	Land, Houses, Index funds
Skills: AI will eventually do better than you	→	Audience (personal brand and trust) Intellectual property (patents, content, trademarks)

It's like with an investment portfolio: You don't want to hold cash because it will evaporate with inflation. You want to turn it into property and stock indexes. To be well off in the future, you need to stop accumulating "skills" and start turning them into audience and intellectual property.

What to Do Now Before It's Too Late?

You need to execute a trade: Convert your skills into a personal brand audience. It could be a YouTube channel, a newsletter, a music band, a comedy show, a popular Twitter/X profile, anything.

Are you good at programming computers? Start a blog about it. Are you good at marketing? Start a local marketing community meetup. Are you into writing? Find a friend with a guitar and make songs together and put them out on YouTube. Are you good at design? Start your own brand.

Think of it as selling a dying currency to buy gold. You convert a volatile asset (coding, marketing, writing, design) into a stable one (an audience that trusts you). You use your current skills to build the reputation that will survive the death of those skills.

Once Upon a Zimbabwe

So, how do you build this brand? How do you gain an audience when the internet is flooded with AI slop? I have two pieces of advice and both of them are strange.

To understand the first, you have to understand the mechanism of the enemy. You have to understand how large language models (LLMs) actually think.

Simplified, AI models are the "average" of all human output. They are probability machines. They predict the next word based on what is *most likely* to come next in all of the books and text from the internet used for training. If you type "Once upon a ...", the AI screams "TIME!" because that is the statistical average.

This means AI is the God of Average. It is incredible at producing predictable content, and it is terrible at producing anything chaotic.

You must use this to your advantage. The value of predictable content is going to be near zero. Therefore, the value of unpredictability or any fresh thought is skyrocketing.

To survive, you must surprise people. You must do the thing that is statistically unlikely. Here's a metaphor for you: If there is a God watching you, do something that would surprise him.

When the world expects "Once upon a time," you go "Once upon a Zimbabwe." Be the glitch in the pattern. Be the variable that the machine couldn't predict. If your writing, your videos, or your business strategy feels "correct" and "standard," you are marking yourself for deletion. Novelty is your currency.

Stop Being Professional

My second piece of advice is even harder for high achievers to swallow.

For the last 20 years, "value" meant "high production quality." It meant polished videos, perfect grammar, and studio lighting.

That era is over. AI can generate 4k videos. AI can write perfect grammar. AI can polish a turd until it blinds you.

When "perfect" becomes cheap, "authentic" becomes expensive. Use your face. Use your personality. Make it candid. Stay away from anything overly polished. If you are making a YouTube channel, don't buy a $4,000 camera. Take your grandpa's old tape camcorder. Shoot it on a 20-year-old iPod Nano (yes, they had cameras).

Why? Because the graininess proves it's real.

The value of content that feels human and authentic will go up because it triggers the "Taylor Swift Effect." It builds connection.

If you're a stand-up comedian, be clumsy. If you're a musician, leave the breath in the recording. If you're a writer, leave the weird turn of phrase.

> Your enemy (AI) will use blandness and professionalism. "Professional" means standardized, standardized means replaceable.
>
> Your allies—the two things that will increase in value in the AI age—are **unpredictability** and **authenticity**.

How Average Dies

So, should you boycott AI? Should you smash the servers like a Luddite?

Absolutely not (unless you wanna be broke).

Don't replace yourself with AI. Enhance yourself with it.

Don't let it write for you, but instead let it give you feedback.

Every hero needs a mentor. But in the digital age, mentors are scarce. You cannot email Elon Musk or Steve Jobs and ask for a critique of your strategy. But here is the twist, and it is the most important distinction you will learn in these pages: If you ask 100 random writers for feedback on your manuscript, they will give you 100 different opinions. If you average those opinions out, you get "the median opinion"—or an opinion of ChatGPT.

Here is the question: Would you prefer to hear the opinion of one genius, or the single, mushy, averaged opinion of "an average person"?

When you ask ChatGPT for business advice, you aren't asking a billionaire. You are asking for the average advice of everyone on the internet. You are asking for "average." And as we've established, average dies.

You don't want the average; you want the outlier. You want the specific bias of a master, or several masters. That is the point of a new category of AI products coming to market: AI based on a specific person, usually with their consent. I am happy to be involved in one of them called Callable.

You don't have to use this specific one, but I think we've built something crazy (if you'd like to try, send me an email and I'll happily give you the free credits). We realized that to build a unique personal brand—to be the "Once Upon a Zimbabwe" in a sea of "Once Upon a Times"—you need feedback from the 1%.

Callable allows users to speak to AI digital twins of famous people in technology, business, nutrition, psychology, and almost any other topic. It doesn't give you the "average" answer. It gives you *their* answer.

I've been personally using it every day. I ask the digital twins for opinions on my everyday things: my diet, my code, my writing, my business. The quality of that opinion has been vastly greater than generic chatbots like ChatGPT or Gemini because it has a point of view. It helps me sharpen my chaos, rather than smoothing it out.

The future still belongs to humans. But only those who use the machines to become *more* unique, not less.

About the Author

Vlad Grankovsky is a computer scientist, entrepreneur, and speaker. Vlad has been a computer science teacher, did machine learning for neuroscience, and did several AI startups, which included building a humanoid robot called Captcha. Currently Vlad works on Callable, a startup allowing experts to scale their impact through digital replicas that the users can talk to.

Email: vlad@callable.com.ai
Website: www.callable.com.ai
LinkedIn: https://www.linkedin.com/in/vladimir-grankovsky/

CHAPTER 15

UNCOMFORTABLE QUESTIONS

By Daniele Grassi
CEO and Co-Founder at Axyon AI
Modena, Italy

The illiterate of the 21st century will not be those who cannot read and write, but those who cannot learn, unlearn, and relearn.

—Alvin Toffler

Almost all of us have experienced that moment: a kind of vertigo that lands on you when, talking to an AI system, you realise that had you not known it was a machine, you would have mistaken it for a person. The reassuring feeling of giving orders to a computer was gone. We were facing something surprisingly coherent, using language that seemed to carry genuine intention. And as we read a response that appeared thought through, a silent, unsettling question crept in: If a machine can do this too, what remains exclusively ours? The unease behind that question has everything to do with identity. For generations, we considered intelligence—the ability to reason and create, to interpret the world—as our sacred temple, the very essence of being human. Today, the doors of that temple have been opened, and what enters resembles us too closely to be dismissed.

The expression "AI edge" is usually understood as the advantage that artificial intelligence holds over humans in synthesis, analysis, computation, and prediction. I think this reading tells us more about ourselves than about any technology. We keep framing every new tool as a challenge to human nature, as if the only interesting question were "Who wins?" I find a different question far more compelling: How must humans evolve to thrive in a world where AI is part of everyday life, much like electricity or the Internet already is? The real edge will belong to the human being who evolves to live and prosper alongside it.

In the financial markets, this idea is especially visible. AI has moved beyond experimentation to become an invisible operational infrastructure. At the deepest levels, algorithms govern order routing and optimise liquidity; higher up, they support investment strategies, thematic assessments, rebalancing, and portfolio analysis. In many areas, they operate in a fully automated way. The human being remains, but increasingly as supervisor and interpreter, setting direction rather than executing every phase of the process. It is a natural transition in complex systems where the volume of data grows at a pace that renders manual management impossible.

I run a company that builds these AI systems and lead a team of over 40 people—physicists, engineers, data scientists, software developers—who push AI to its limits every day. So when I say this technology changes the nature of work, I am speaking from the inside.

While the efficiency of automated execution is now widely accepted, genuine comprehension of the process still raises resistance. When discussing artificial intelligence, a recurring theme surfaces: the demand for "explainability." AI models are asked to justify their decisions as if they were rational, transparent subjects. I hear this in almost every client meeting. Someone will ask: "Can you explain why the model allocated 12% to this sector?" Fair question. But when I ask the same person to explain why they made their last major hiring decision—to trace the actual reasoning, step by step—the room goes quiet. This demand presupposes something we rarely examine: the assumption that human beings themselves are explainable.

Psychology and neuroscience have long shown that most of our decisions do not follow a conscious, linear process. We construct coherent narratives after the fact, stories that reinterpret rather than describe what actually happened. Once you see this, the opposition between "human explainability" and "artificial opacity" starts to crumble. A model's decision is deterministic, traceable to a very long equation that will always yield the same result; a human decision is not traceable even in principle. The transparency we actually need is about the AI's world: what data it sees, what objectives it has been assigned, what ethical and operational vision it embodies, and what biases we have inadvertently built into it.

The operational impact of AI goes beyond its predictive edge. What changes the game is the sheer magnitude of the acceleration. Tasks that required months of cerebral effort by individuals can now be completed in hours, or even minutes. In the world of investments, evaluating hundreds of financial instruments, constructing complex decision-making frameworks, analysing heterogeneous data sources, and stress-testing assumptions—these are increasingly the output of automated processes delivering unprecedented quality and speed. The change is qualitative. When the scale of time shifts, the content of work shifts with it: What was previously too costly to do becomes routine. And what was considered routine suddenly becomes insufficient.

This compression of time does something to people. I have watched it across my own organisation and across the industry. The most visible layer is the fear of obsolescence. For centuries, automation threatened physical labour, the repetitive and the mechanical; we felt safe, sheltered behind our intellect. Today, for the first time, automation touches cognitive work: writing, analysing, designing, deciding. If you have built your social identity on competence, and software performs that competence in three seconds, the doubt moves past economics into something existential. Am I still needed? Even those who do not fear for their jobs feel perpetually behind, behind on the world itself, behind every week's "revolutionary" new tool. A race with no finish line, producing only new starting points. And deeper still, AI questions the story we have told ourselves since the dawn of culture. We defined ourselves through our creative and linguistic intelligence. These capabilities were proof of our uniqueness, and now we watch them being reproduced, perhaps without consciousness, but convincingly enough to make us doubt.

Creativity is where this gets personal for me. I have been programming since I was eight years old, on a Commodore 64, in 1992. Every language I have learned—BASIC, Pascal, C, Java, Python—is a scar from a problem I fought and eventually solved. When I asked my own team what scared them most about AI in their work, nobody said "losing my job." They said "losing the pleasure of programming." That made perfect sense.

There is a deep satisfaction in wrestling with a problem, in the friction of building something yourself. Generative models can transform any human creative spark into text, images, code, and strategies. They offer unprecedented expressive power, and the cost is one that most people miss until it is too late. Think of Google Maps: Practically no one today relies on their own sense of direction, and the loss becomes evident only when you try to exercise it again. With AI, the risk is broader: Cognitive delegation concerns the entire creative and decision-making process. If delegation becomes total, it leads to atrophy: We become incapable of facing complexity without external tools, and we are precisely fragile where we should be strongest. The person who surrenders their own cognitive autonomy is the one we should actually worry about. That is why it becomes essential to preserve spaces for unassisted creativity, spaces where we think and decide for ourselves. Unmediated. Messy. I still write code by hand sometimes, for no practical reason, just to feel the resistance of the problem.

The risks of cognitive offloading extend well beyond individual skill loss. We are witnessing what some theorists call the "dead internet theory": a scenario in which the web ceases to be a forum for human interaction and becomes a corridor of echoes between bots, algorithms that write content to be read by other algorithms. A study published in *Nature* has formalised the concept of "model collapse": When AI feeds on data generated by other AI in a recursive training cycle, information quality degrades irreversibly. I think of it as digital inbreeding: the data becomes increasingly amorphous, stereotyped, detached from reality. When the training data itself is synthetic, the model drifts further from anything that was ever real. The implications for any field that depends on reliable information—from journalism to science to the financial markets where I work—are hard to overstate. We risk an entropy of truth in which authentic human knowledge—complex and disorderly—is overwritten by a sterile statistical mean.

In this context, the most important capability is maintaining your own cognitive competencies while learning to structure decisions in a world where the collection, aggregation, synthesis, and interpretation of information are entrusted to models. The human being must learn to design the decision-making process, to define which data matter and which do not, which questions deserve attention, and which ethical and moral criteria should guide the evaluation. AI does not free us from responsibility. It shifts it higher, to a place where many are not accustomed to operating. The person who succeeds will be the one who can hold the big picture, the one who understands enough about the tools to know what to ask and enough about the world to know which answers matter.

This new role runs into a powerful barrier: fear. Fear of being surpassed, of losing relevance. I have seen it in portfolio managers as much as in software developers — often precisely in those who would have the most tools to integrate AI into their work. And I do not mean the abstract kind. I mean the kind that makes a senior developer avoid a tool they know would help, because using it means admitting that something they spent 15 years mastering can now be done in seconds. In an age where the pace of technological change is unprecedented, standing still is the riskiest move you can make.

There is a way through this, I think, that avoids both uncritical enthusiasm and nostalgic rejection. I would call it a technological *amor fati*—a love of fate. The phrase belongs to Nietzsche, but its roots are Stoic. Epictetus taught that freedom begins in distinguishing what depends on us—our choices, our judgments—from what does not. The AI revolution does not depend on you. Your response to it does. This is about recognising the playing field you are on, so you can act with intention rather than panic. Rejecting reality would be like rejecting the rain: You can be as angry as you want, but you will still get wet. Only the person who accepts that it is raining can decide to open an umbrella.

I have learned this at Axyon AI, too: AI adoption is cultural, and I say that as someone who runs an AI company. When we introduced new models into our own workflows, the technical deployment was the easy part. The hard part was the team. People who had spent years building expertise in a particular analytical approach did not want to hand it over to an algorithm, even one they had helped build. I have seen the same dynamic at client organisations that spend millions on AI infrastructure and then watch it gather dust because nobody addressed the fear in the room. The question worth asking is always the same: How would we have designed this process if AI had existed from the beginning?

All of this leads me to a belief I have held for years, reinforced every day by building AI systems and watching what they do to the people around them: AI will not replace humans. That much I am sure of. It will replace humans who refuse to evolve. The real advantage is human: curiosity, the willingness to think with clarity under pressure, the patience to sit with ambiguity, and the stubborn refusal to outsource your own judgment.

In a world that races forward at furious speed, the most valuable skill is steadiness. The kind of calm that comes from standing inside the storm with your eyes open. No algorithm, however advanced, will ever carry the weight of a real decision or share a silence with someone who is afraid. No system will refuse an optimisation because it crosses a line. These are things only a human being can do: Be present when it matters and carry the full weight of a choice, including the weight of saying no when every number says yes. AI is forcing us to figure out what we consider truly special about ourselves. The question is uncomfortable. Good. We may have been avoiding it for too long.

About the Author

Daniele Grassi is the founder and CEO of Axyon AI, a technology company that has been developing frontier artificial intelligence systems for the financial industry for over a decade. Leading a team of more than 40 professionals—physicists, data scientists, and software engineers—he drives the design and deployment of AI-powered investment solutions. Daniele began programming at the age of eight on a Commodore 64 and has since built a career at the intersection of technology and finance. His work lives at the frontier of AI innovation, while his personal reflections explore the human dimension of this transformation: how individuals, organisations, and societies can evolve alongside intelligent machines without losing what makes them fundamentally human.

Email: daniele.grassi@axyon.ai

Website: www.axyon.ai

LinkedIn: https://www.linkedin.com/in/danielegrassi84/

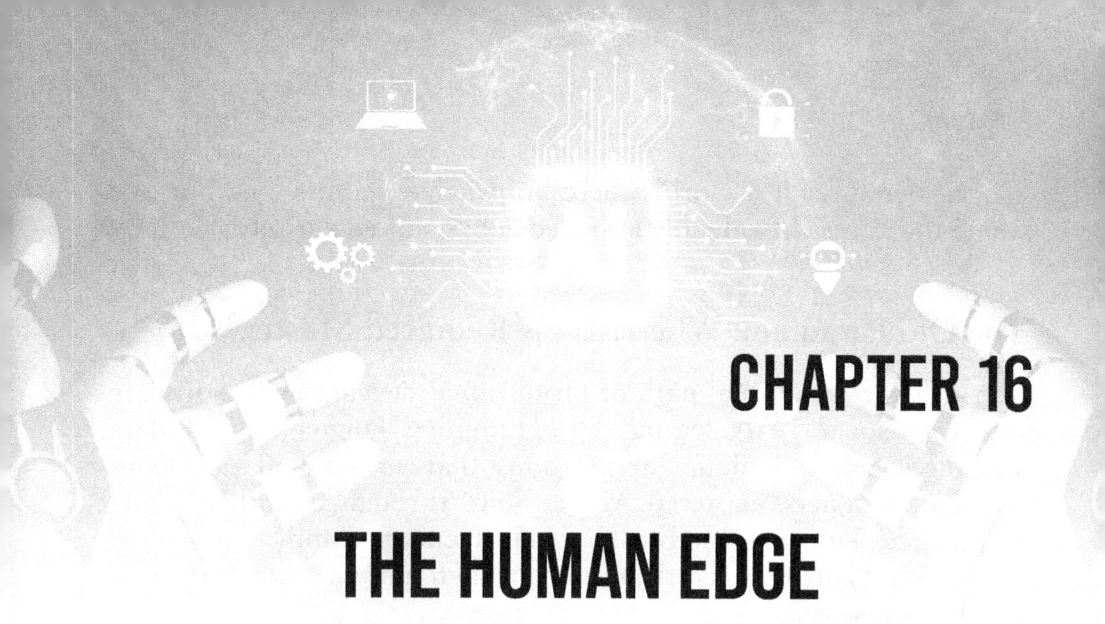

CHAPTER 16

THE HUMAN EDGE

By Prince Jain
AI Technology Leader
San Jose, California

The future is already here—it's just not evenly distributed yet.
—William Gibson

When I started in technology nearly a decade ago, the path to meaningful contribution was clearly defined: Invest years mastering technical fundamentals, accumulate expertise through repetition, and eventually leverage that knowledge to create something valuable. This was the covenant of the knowledge economy: Spend thousands of hours learning syntax, memorizing frameworks, and perfecting execution skills, and you'd earn the right to build.

That covenant is breaking. Not gradually, but rapidly.

Artificial intelligence hasn't invalidated expertise, but it has fundamentally compressed the timeline between conception and execution. The traditional barriers, technical skill gaps, time constraints, and resource limitations are dissolving at an unprecedented pace.

What emerges isn't a world where knowledge matters less, but one where the nature of valuable knowledge has shifted entirely.

The Old Paradigm: When Access Required Mastery

Consider the traditional path of professional development. Software engineers spent years learning programming languages and data structures before building applications that solved real problems. Graphic designers mastered Adobe Suite through countless hours of practice. Financial analysts became fluent in complex modeling techniques. Writers honed their craft through innumerable drafts and rejections.

This accumulation model created natural gatekeepers. Only those with sufficient time, resources, and persistence could participate in creation at scale. Innovation was necessarily slow because each innovator had to rebuild foundational knowledge from scratch.

But here's what we're discovering: Much of what we spent years memorizing wasn't about deep understanding, it was about access to execution. We learned syntax because computers couldn't understand natural language. We memorized formulas because calculation was manual. We perfected techniques because execution required trained hands. The knowledge itself often wasn't the barrier to creativity; the tools were.

Working at companies like Google and Meta, I watched this pattern repeat across domains. So much of what we celebrated as "expertise" was really sophisticated workaround development for poorly designed interfaces between human intention and machine execution.

The Disruption: From "How" to "What" and "Why"

AI technology represents a fundamental shift in what humans need to know versus what machines can handle. When AI can generate code from natural language descriptions, write initial drafts from outlines, or create visualizations from verbal instructions, the bottleneck moves. The critical question shifts from "Can I execute this idea?" to "Is this idea worth executing?"

This compression doesn't make us learn new technology so much as it makes us more capable within whatever domain we already occupy. A marketing professional with 20 years of consumer behavior expertise no longer needs five additional years learning to code to build a customer analytics tool. They can describe what they need, and AI can help them build it. Their accumulated wisdom about customer psychology suddenly becomes directly actionable.

I see this transformation accelerating. A finance executive who previously required a three-person development team can now prototype forecasting tools over a weekend. A healthcare administrator who's spent years navigating insurance complexity can build claims processing assistants without writing code. The constraint has fundamentally shifted.

The move from "how" to "what" and "why" transforms everyone into potential innovators and creators. When execution barriers drop, the differentiator becomes vision. Can you identify the right problem? Do you understand the true need? Can you imagine a better solution? These questions require domain expertise, but expertise of a different kind, contextual understanding rather than technical memorization.

This is particularly transformative for executives and industry veterans who have spent decades accumulating knowledge about market dynamics, customer needs, organizational behavior, and industry nuances. Previously, their insights required intermediaries, teams of specialists to translate vision into reality. Now that translation layer is compressing. A CEO who deeply understands their customer base can rapidly prototype solutions, test assumptions, and iterate without assembling entire departments first.

The Coming Wave: The Age of Experimentation (Next 5–10 Years)

If we project forward, the immediate future belongs to experimentation. With technical barriers lowered, success depends on market understanding: What segment should we serve? What do customers truly value? What problems are we actually solving?

This experimental phase will be characterized by rapid prototyping and ruthless iteration. Where building a minimum viable product once took months or years, AI-assisted development compresses that timeline to weeks or days. When the cost of being wrong drops by an order of magnitude, strategy changes completely. You don't spend six months planning the perfect product. You build three versions in a month, see which one customers actually use, and pivot accordingly.

Successful individuals and organizations in this era will be distinguished not by execution ability (because execution becomes increasingly democratized) but by their capacity to ask insightful questions and interpret results. They'll build fast, learn faster, and pivot before competitors launch their first attempt. Competitive advantage lies in velocity of learning rather than perfection of craft.

This creates remarkable opportunities for revolutionizing established industries. A physician who has spent 30 years observing patient outcomes and treatment responses can now rapidly develop and test decision-support tools without becoming a software engineer. A teacher who understands how students actually learn can prototype adaptive learning systems without waiting for EdTech companies to build what they imagine might work. The people closest to problems can now build solutions directly.

The experimental era also democratizes innovation in ways we're only beginning to appreciate. Startups can compete with established players not by matching their resources but by out-learning them. Small teams can test more hypotheses in a month than large organizations previously could in a year. The question "What should we build?" matters infinitely more than "Can we build it?"

I've watched this pattern emerge in real time. Companies that embrace rapid experimentation are iterating through more hypotheses in a quarter than their competitors test in a year, not because they're smarter, but because they're faster and less precious about their initial assumptions.

But experimentation alone isn't the end state. It's a transitional phase, necessary, transformative, but ultimately giving way to something deeper.

The Horizon: The Age of Exploration (15–20 Years)

As AI capabilities mature and the experimental phase yields its lessons, we'll transition into what might be called the Age of Exploration. This isn't about finding new markets or optimizing existing solutions. It's about asking "why" at a fundamental level.

Why do we organize industries this way? Why do we assume certain limitations? Why do we accept particular trade-offs that made sense when execution was expensive but may no longer hold?

The exploration phase will be characterized by reconceptualization and the pursuit of previously impossible goals. When execution becomes nearly frictionless and we've experimented our way through obvious applications, attention turns to reimagining entire systems. We won't just ask, "How can AI help us do X better?" but "Should we be doing X at all?" or "What becomes possible if we remove constraint Y entirely?"

Consider healthcare as an example. We currently organize around scarcity. Doctor time is limited, so we triage, batch appointments, and accept that most people can't access care when they need it. But what if diagnostic expertise becomes abundant? Do we still need the same hospital structures? The same insurance models? The same regulatory frameworks designed around scarcity?

These aren't hypotheticals I'm comfortable answering definitively. Neither is anyone else yet. But these are the questions the exploration phase will force us to confront.

This is where sustainability, in every sense, becomes central. Not just environmental sustainability, though that's crucial, but sustainable business models, growth patterns, and social structures. With the ability to build and iterate rapidly, we can finally address second-order and third-order effects of our innovations. We can design for long-term flourishing rather than short-term optimization.

The exploration age will likely see massive industry consolidation around fundamentally new paradigms. Just as the internet didn't merely make existing businesses more efficient but enabled entirely new structures, social networks, cloud computing, and the gig economy, mature AI will enable organizational and economic models we haven't yet imagined.

This is where domain expertise becomes even more valuable, but in a different dimension. Understanding not just how an industry works but why it works that way, and whether it should, requires deep historical knowledge, systems thinking, and the wisdom to recognize unintended consequences. The professionals who thrive in this era will be those who maintained their curiosity and expanded their thinking beyond pure optimization.

The Power of Strategic Reasoning

What makes this transition particularly significant is that it rewards different capabilities at different stages. The experimental phase favors speed, adaptability, and market intuition. The exploration phase favors depth, synthesis, and ethical reasoning. Organizations and individuals who recognize which phase they're in and what that phase demands will have an enormous advantage.

For those currently building careers, this suggests a strategy that might seem counterintuitive: Don't just learn AI tools, though that's valuable. More importantly, deepen your understanding of your domain. Become the person who understands customer behavior so intuitively you can predict it. Study the historical evolution of your industry until you see patterns others miss. Build relationships and develop judgment about human needs and organizational dynamics.

The technical skills, the "how," will continue to become more accessible. But understanding what to build and why to build it requires something AI cannot provide: lived experience, accumulated wisdom, and deep contextual understanding. As AI handles more execution, the value of human judgment, creativity, and strategic insight increases rather than decreases.

Making Everyone a Creator

Perhaps the most profound implication of this shift is the democratization of creation itself. When the barrier to bringing ideas to life drops dramatically, we discover something surprising: Most people have valuable ideas. They were simply never able to act on them because the execution cost was prohibitive.

A teacher with insights about learning. A nurse with ideas for patient care. A retail worker who understands customer behavior. A parent managing household logistics. All of these people observe problems and imagine solutions daily but traditionally lacked the technical means to build those solutions.

This democratization doesn't diminish expertise, it amplifies it. The teacher doesn't need less pedagogical knowledge; they need more, because now they can actually build the adaptive learning tool they've always imagined. The nurse's decades of patient care experience become directly translatable into better healthcare systems. Domain expertise becomes buildable expertise.

We're moving toward a world where the limiting factor isn't "Who can build?" but "Who understands what's worth building?" That represents a fundamental shift. It suggests that the next wave of innovation will come not exclusively from technical prodigies but from domain experts across every field who finally have the tools to make their insights real.

The Challenge Ahead

This transformation isn't without challenges. As creation becomes democratized, so does the potential for poorly conceived solutions. When everyone can build, not everyone will build well. The experimental phase will be messy, littered with failed ventures and misguided applications. The exploration phase will force difficult conversations about values, priorities, and the kind of future we want to create.

If you've built your career on being the person who can execute, the developer who knows all the frameworks, the designer who's mastered all the tools, the analyst who's fluent in every platform, your competitive advantage is shifting. Not disappearing, but evolving. The shift is from execution to judgment, from knowing how to do things to understanding what's worth doing and why.

But these are ultimately problems of abundance rather than scarcity. The risk of moving too fast is preferable to the tragedy of moving too slowly. The danger of too many experiments is better than the stagnation of too few. And the difficulty of choosing between many possible futures is infinitely preferable to having no agency over the future at all.

Conclusion: The Acceleration Is Just Beginning

We stand at the beginning of this great acceleration. The compression of timelines, the elimination of barriers, the shift from accumulation to experimentation to exploration, these aren't distant possibilities but current realities unfolding around us.

The professionals and organizations that recognize this shift and adapt their strategies accordingly will find themselves with unprecedented power to shape their industries and solve previously intractable problems. The key insight is this: AI advancement isn't primarily about learning new technology for technology's sake. It's about becoming more capable and powerful in whatever domain you already occupy. It's about shifting focus from the mechanics of execution to the substance of creation. It's about asking better questions because we finally have the tools to act on the answers.

The next few years will reward experimentation, rapid testing, quick learning, bold attempts. The following decades will reward exploration, deep thinking, ethical reasoning, and the courage to reimagine entire systems. But both phases reward the same fundamental quality: the wisdom to know what's worth building and why.

For those willing to embrace this shift, the opportunity is remarkable: not just to build better solutions, but to become the kind of person who can envision a better future and actually bring it into being. That's the promise of AI, not that it replaces human creativity, but that it finally allows human creativity to operate at its full potential, unencumbered by the barriers that have historically kept ideas trapped as dreams rather than realized as reality.

The timeline is shrinking. The barriers are falling. The question now is: What will you create?

About the Author

Prince Jain is an AI technology leader based in the Bay Area, California. With nearly a decade of experience building scalable AI solutions at leading technology companies including Google and Meta, Prince has been at the forefront of the AI revolution, witnessing firsthand how artificial intelligence is eliminating traditional barriers between conception and creation.

Throughout his career, Prince has solved some of the most challenging problems in AI infrastructure, from optimizing compilers and runtime systems to enabling production models that serve billions of users daily. His work has consistently delivered breakthrough performance improvements and accelerated AI deployment across multiple platforms, bridging the gap between cutting-edge research and real-world applications.

Educated at Carnegie Mellon University, Harvard University, and the Indian Institute of Technology (IIT) Roorkee, Prince combines deep technical expertise with strategic business acumen. His studies in business consulting, finance, leadership, and monetary policy at Harvard have given him a unique perspective on AI's evolution from both technological and market dynamics standpoints.

His perspective on AI's transformative potential comes not from theoretical speculation but from daily engagement with the tools and systems that are reshaping how we build, create, and innovate. Prince writes from the front lines of the AI revolution, where the shift from "how" to "what" and "why" isn't a future prediction but a present reality.

Email: princejain1101@gmail.com
LinkedIn: https://www.linkedin.com/in/prince-jain-63646580/

CHAPTER 17

NAVIGATING UNCERTAIN WATERS: LEADERSHIP LESSONS FOR THE AI ERA

By Todd S. James
CEO, Fortune 25 AI & Tech Executive
Cincinnati, Ohio

It seems to be a law of nature, inflexible and inexorable, that those who will not risk cannot win.

—John Paul Jones

Risk does not announce itself politely. At sea, it arrives in shifting winds, rising swells, and decisions that cannot be deferred. There is rarely enough information, never enough time, and always a consequence waiting on the other side of action or inaction.

That reality defined my earliest years as an officer in the United States Coast Guard. While my service took me across the world, my most formative time was spent in and around a stretch of water known as the Graveyard of the Pacific. It is a place of abundance and danger, a major maritime corridor shaped by unpredictable weather, powerful tides, and unforgiving rocky shorelines.

More than two thousand ships have been lost there. Many mariners never returned home. In that environment, risk is not theoretical or retrospective. It is immediate. It strips decisions down to their essentials: incomplete information, limited time, and outcomes that cannot be reversed. Moving too slowly can be just as dangerous as moving too fast. The hardest part is knowing the difference.

It was there, long before boardrooms and balance sheets, that I learned what responsible risk truly means. Bold action can be rewarded, but only when grounded in preparation, judgment, and teamwork. Miscalculation, overconfidence, or poor timing carry consequences that are swift and unforgiving. Those lessons followed me well beyond the service, shaping how I think about leadership, uncertainty, and decision-making in every environment since.

AI Has Changed the Risk Equation

Years later, I began to recognize the same dynamics in a very different setting. The uniforms were replaced by suits, and increasingly by t-shirts and hoodies. The ocean made way for conference rooms. But the conditions felt familiar. Senior leaders were navigating fast-changing environments, incomplete information, and decisions with real consequences. Artificial intelligence entered the picture not as a single tool, but as a force reshaping how work gets done, how value is created, and how competitive advantage is gained or lost.

In conversations with CEOs and boards, I hear a consistent tension. Leaders worry about moving too fast, investing in the wrong capabilities, or introducing risks they do not fully understand. At the same time, they feel pressure to act. Competitors are advancing. Expectations are shifting. Markets are not waiting.

What makes this moment particularly challenging is that AI changes the nature of risk itself. Decision cycles compress. Information moves faster than traditional governance and operating models were designed to handle. The danger is often not recklessness, but misalignment. Not ambition, but confusion. Many organizations sense the urgency, but lack a clear course forward.

I see this play out in organizations with technically working AI models sitting on the sidelines. In one financial services company, a team developed models to identify which customers of competing firms were most likely to switch in a specific line of business. The analysis was sound. The models performed as expected.

What followed was not a technical debate, but a prolonged hesitation around commitment to deploy and organizational acceptance of the model's recommendations. Leadership revisited edge cases and governance questions long after the pilot had proven viable. By the time a decision was finally made, the business conditions that made the use case compelling had shifted, and the business was exited altogether.

Internally, it was summed up bluntly: The surgery was successful, but the patient was dead. The technology worked. The moment was gone.

The False Choice Leaders Feel

In response, many leaders frame AI as a trade-off between two risks. Move aggressively and risk wasted capital, regulatory exposure, or reputational damage. Move cautiously and risk falling behind competitors who are already learning, adapting, and scaling.

It feels like a rational choice. In reality, it is a false one.

We can see the consequences of both extremes. Organizations that hesitate are already losing ground as AI leaders pull away, compounding advantages in productivity, insight, and execution. At the same time, highly visible failures show what happens when AI is deployed carelessly, without discipline or oversight.

The lesson is not that one side is right and the other wrong. Risk is not defined by speed or caution, but by how well leaders read the situation and commit to a course.

At sea, the most dangerous moments were rarely about bold action or restraint alone. They came when leaders misread conditions, delayed decisions too long, or acted without preparation. Waiting for certainty was not caution. It was exposure. Acting without judgment was no better.

AI presents the same challenge. It does not reward recklessness, but it punishes indecision. The leaders who succeed are not choosing speed over safety or caution over progress. They are learning how to move deliberately in an uncertain environment. In the AI era, the greatest leadership risk is not moving too fast or too slow, but misunderstanding which one you're actually doing.

How Leaders Navigate Uncertainty

In situations like these, the question is not how to eliminate risk. That is never possible. The question is how leaders navigate it without becoming reckless or frozen. Over time, both at sea and later in business, I found myself returning to three disciplines. They are not formulas. They are ways of thinking, based on formative experience, that hold up when conditions are unclear and the stakes are real.

The first is making the best decision possible with the information available. Put simply, endeavor to execute sound judgement. At sea, waiting for perfect data was rarely an option. Conditions changed faster than forecasts. Decisions had to be made knowing they might need to be adjusted minutes later. The goal was not certainty, but direction. The same applies to AI. Leaders who wait until every risk is resolved often find the environment has already shifted beneath them. Progress comes from acting with intent, learning quickly, and refining decisions as new information emerges.

The second discipline is falling back on skill, preparation, and operational discipline. In the Coast Guard, training mattered most when plans broke down. In business, this translates to building foundational capability before chasing scale. Leaders who succeed with AI invest early in principles, clear decision rights, operating models, and governance designed to enable, not contain. Not to slow teams down, but to give them a stable platform from which to move faster without creating chaos.

The third is relying on the strength of the team around you, your network. No meaningful operation at sea is carried out alone. Navigation, engineering, communications, and command must work in sync, often in coordination with a network of other assets. AI demands the same coordination. Organizations that make progress break down silos between business, technology, legal, and operations. They bring boards along early. Risk becomes manageable when responsibility is shared and aligned.

Together, these disciplines address the false choice leaders often feel. They allow organizations to move without rushing and to learn without drifting. Risk remains, but it becomes something leaders can work with rather than fear. Risk does not disappear with caution. It accumulates when decisions are delayed.

Patterns From Companies That Get It Right

Across industries, the organizations that make real progress with AI share a small number of patterns. They are not defined by superior algorithms or larger budgets. They succeed because leaders apply discipline to how risk is taken, how decisions are made, and how learning compounds over time. They understand that governance, when properly crafted, is the bridge between innovation and value at scale.

First, they are clear about where AI is expected to create value. Successful leaders resist chasing use cases simply because the technology is available. They anchor AI efforts to specific outcomes such as revenue growth, margin improvement, customer experience, or operational resilience. They measure and track it relentlessly. This clarity reduces noise and gives teams a basis for deciding what to accelerate and what to stop.

Second, they sequence intentionally. These organizations do not attempt to transform everything at once. They start where data is strong, decision velocity is high, and impact is visible. Early wins build confidence and provide a basis for reenvisioning operating models and strengthening governance to enable organizations to unlock expanded ambitions. Over time, these early wins establish reusable capabilities that increase speed and compound value across future efforts.

Third, they design governance to enable movement rather than prevent it. In struggling organizations, governance becomes synonymous with delay. In effective ones, it provides guardrails that allow teams to act with confidence and greater levels of autonomy. Risk is addressed upfront and in-line rather than rediscovered late.

Finally, leadership engagement goes beyond sponsorship. Senior leaders stay close to the work. They ask hard questions, resolve tensions across functions, and signal that AI is core to how the business operates, not a side project. The same is true for the board, where AI-fluency is emerging as an essential competency to guide organizations and fulfill fiduciary responsibilities.

These patterns do not eliminate uncertainty. They make it productive.

The Real Risk Is Misunderstanding Risk

Many organizations misinterpret AI risk. Risk is treated as something to avoid rather than manage deliberately. Governance is viewed as a brake rather than a steering mechanism. The result is hesitation disguised as responsibility.

The most visible failures in AI are rarely cautionary tales about moving too fast. They are reminders of what happens when AI is deployed without clarity, accountability, or oversight. The greater risk for most organizations is not recklessness. It is paralysis.

Effective leaders separate ethical and regulatory concerns from emotional discomfort. They define where AI is allowed to act, where human judgment must remain in control, and how outcomes will be monitored. Governance becomes a tool for learning rather than delay, the bridge between innovation and scale. Transparency replaces guesswork. Accountability replaces anxiety.

At sea, risk was never eliminated. It was made visible, named, and managed. That discipline did not slow us down. It allowed us to move with confidence when others hesitated. The same is true in business.

We have seen this pattern before. Kodak invented the digital camera but protected film. Blockbuster optimized stores while streaming rewired consumer behavior. Nokia dominated handsets while losing the software platform. In each case, leaders did not ignore change. They studied it, piloted it, and governed it, until governance became containment and the market made the decision for them. The parallel to AI governance today is direct.

The failure was not a lack of intelligence or effort. It was a misreading of where the real risk sat as the curve was bending.

In a similar vein, this is why so many organizations can point to AI activity yet struggle to demonstrate meaningful business value.

A Practical Course Forward

For leaders navigating AI, the course forward does not begin with certainty. It begins with clarity. Clarity about where AI can realistically create value. Clarity about which risks matter most. And clarity about who is accountable for decisions as learning unfolds.

The first step is intent over urgency. Leaders who make progress resist the pressure to act broadly and instead choose a small number of priorities where better decisions or faster execution would materially matter. Early AI efforts are treated as learning vehicles, not final answers. This focus reduces noise, sharpens accountability, and prevents activity from outpacing understanding.

The second step is investing in foundations before scale. As earlier efforts revealed, most failures do not come from experimentation itself, but from attempting to scale what was never designed to endure. That work is both organizational and technical. Clear decision rights, simple governance, and shared language reduce friction as use cases expand. At the same time, successful teams build technology for reuse instead of one-off solutions. Data pipelines, model components, and integration patterns are designed once and extended many times, allowing progress to compound rather than reset.

The third step is visible leadership. AI initiatives stall when leaders step back too soon, assuming momentum will sustain itself. In practice, sustained progress requires continued engagement. Leaders stay involved beyond the launch, reinforce priorities as trade-offs emerge, and bring boards into the conversation early. Oversight becomes a source of alignment rather than delay, keeping responsibility and speed in balance.

This course does not eliminate risk. It replaces fear with judgment. And judgment, consistently applied, is what allows organizations to move forward even when the horizon is unclear.

Learning to Navigate

At sea, uncertainty never disappeared. Conditions shifted. Information lagged reality. What mattered was learning how to navigate, adjusting course as conditions changed and acting with discipline when it mattered most.

AI now places leaders in that same position. The technology is powerful. The pace is accelerating. Standing still is no longer neutral or safe. But the answer is not blind speed or rigid control. It is informed action.

The organizations that thrive in this next era will treat AI not as a technical upgrade or a risk to be feared, but as a leadership challenge. One that demands clarity of intent, strength of execution, and the willingness to adjust course when the sea changes.

The waters ahead are uncertain. They always have been. The decision for leaders is how they will navigate.

About the Author

Todd S. James is the founder and CEO of Aurora Insights, where he advises CEOs, boards, and executive teams on how to translate artificial intelligence into durable business value. A former Fortune-25 executive, Todd led enterprise-wide AI strategy at the Kroger Co. and served as chief data and technology officer of 84.51°, Kroger's data science and retail media subsidiary, embedding AI into core commercial and operational decisions at scale.

Earlier in his career, he held senior leadership roles at Fidelity Investments and Deloitte and served as an officer in the United States Coast Guard, experiences that shaped his approach to leadership, risk, and decision-making under uncertainty.

Todd is a frequent speaker on AI leadership and governance, and his ideas have been cited by *Harvard Business Review*, the *Harvard Data Science Review*, *Forbes*, and other executive and industry outlets.

Email: tjames@aurorainsightsllc.com

Website: https://www.aurorainsightsllc.com/

CHAPTER 18

OPERATIONAL AI AND THE RISE OF THE MID-MARKET

By Alejandro Laplana
CEO at Shokworks
Austin, Texas

There are moments in economic history when a new technology emerges not as an incremental improvement but as a force that alters the very foundations of competition. Electricity did this when it transformed manufacturing from muscle-driven to machine-driven operations. Computing did this when information became both instantaneous and scalable. The internet did this when geography collapsed as a barrier to knowledge and commerce. And now, artificial intelligence is reshaping competition in a way that extends beyond past transformations. AI does not merely automate tasks or enhance productivity. It reconfigures the logic of decision-making, coordination, value creation, and organizational structure. It changes what firms can do, how they operate, and ultimately, what determines leadership.

For decades, mid-market firms—those companies between $20 million and $500 million in revenue—have lived with an implicit ceiling. They lacked the armies of analysts, the redundancy of large teams, and the capital reserves that allowed Fortune 500 incumbents to absorb inefficiency. They competed through ingenuity, expertise, and customer intimacy, but the scale advantages of larger firms appeared insurmountable. Growth often meant greater complexity, heavier coordination burdens, and the need to build structures that undermined their agility. Their strengths were real, but their constraints were structural.

Artificial intelligence has broken these constraints.

Operational AI—AI embedded into everyday workflows, decisions, communications, and customer interactions—has overturned the very logic that once favored incumbents. Traditional advantages of size, such as hierarchical coordination and large-scale specialization, now act as sources of friction. AI automates coordination, improves judgment at scale, and reduces the managerial layers historically required to manage complexity. It embeds intelligence directly into operations, enabling organizations with modest headcount to execute with the precision, responsiveness, and sophistication once accessible only to vast enterprises. At the same time, AI magnifies the advantages mid-market firms always had: agility, alignment, speed of execution, closeness to customers, and a willingness to redesign processes when needed.

This is not a small shift. It represents a profound reordering of competitive dynamics, one that gives mid-market firms a rare asymmetric advantage. But like all inflection points in technology adoption, the opportunity is not permanent. AI follows a diffusion curve: Early adopters gain disproportionate benefits as models learn, data accumulates, workflows evolve, and customers gravitate toward the firms that respond fastest and operate most intelligently. The gap between early adopters and late movers widens not linearly, but exponentially. Those who move now gain a compounding advantage. Those who delay face a permanent cost disadvantage and a strategic position from which catching up becomes prohibitively expensive.

To understand why mid-market firms are so uniquely positioned for leadership in this new era, we must delve into the deeper economic and strategic forces at work, forces explained by the theories of disruptive innovation, creative destruction, and jobs-to-be-done. These frameworks, combined with modern economic analysis of scale, coordination costs, and technological diffusion, explain why AI unlocks advantages for mid-sized firms that large enterprises cannot easily replicate. From there, real-world examples illustrate how mid-market firms are already operationalizing AI to achieve breakthroughs once out of reach. Finally, at the macro level, we consider why mid-market AI adoption will not reduce employment but increase it, driving national productivity and long-term GDP growth.

The conclusion is clear: The mid-market is standing at the threshold of a generational opportunity. Whether they seize it is now a matter of leadership, urgency, and imagination.

The Mid-Market Advantage in the AI Era

To understand why AI uniquely favors mid-market firms, we must begin with the economic logic of how AI alters the structure and purpose of the firm itself. Historically, firms grew larger because scale conferred advantages: They could spread fixed costs across greater volume, build specialized departments, and manage complexity through layers of management. Ronald Coase described companies as coordination mechanisms, structures that reduce transaction costs compared to operating purely through markets. But as organizations grow, their coordination costs rise: Communication bottlenecks, bureaucratic inertia, and cultural fragmentation begin to dilute the benefits of scale. Oliver Williamson later extended this logic, showing that the boundaries of the firm depend on where coordination is cheaper: inside the hierarchy or through the market.

AI profoundly shifts this calculus. Intelligent systems reduce coordination costs by automating data flow, predicting needs, triaging work, and enabling decisions to be made at lower levels with greater accuracy. Tasks that once required managerial oversight can now be handled by AI-augmented employees. Processes that once demanded layers of review can now operate with built-in safeguards and automated analysis. The very reason many firms grew large—because size allowed them to process complexity—becomes less relevant. AI shrinks the minimum efficient scale of many operations, allowing mid-market firms to achieve enterprise-level capability without enterprise-level headcount.

This shift alone gives mid-market companies an opening. But their advantage becomes even more pronounced when viewed through the lens of disruptive innovation, as articulated by Clayton Christensen. Disruptive innovation explains why incumbents often fail to adopt transformative technologies: Doing so requires cannibalizing profitable existing models or challenging entrenched organizational structures. AI threatens billable hours in professional services, established workflows in supply chains, traditional staffing models in customer service, and analysis done by teams of specialists. Large enterprises hesitate because adopting AI undermines the very systems that justify their size. Their legacy revenue models are tied to operational methods that AI would make obsolete. By contrast, mid-market firms are less constrained by legacy business models and more motivated to adopt efficiency-enhancing tools. They face fewer internal barriers, fewer political constraints, and fewer incentives to protect outdated ways of working. Thus, they move earlier, learn faster, and capture value sooner.

A second theoretical lens, creative destruction, reveals even more. Joseph Schumpeter described capitalism as a cycle of constant renewal, where old structures collapse and new ones emerge through innovation. AI is accelerating this cycle dramatically. The long-standing competitive moats built through scale, headcount, and diversified product lines are eroding. Productivity gains no longer require armies of specialists; they require intelligent systems capable of automating routine decisions and elevating human expertise. In this environment, incumbents experience the destructive side of the cycle—their historical advantages lose relevance—while mid-market firms gain access to new forms of productivity that were once unattainable. They do not need to tear down massive structures to rebuild; they can layer AI atop more modern architectures and redesign their operating models with far less friction.

The third lens, jobs-to-be-done (JTBD), further clarifies why mid-market firms are advantaged. JTBD theory argues that customers "hire" products or services to accomplish specific outcomes, and companies that understand these jobs most intimately can innovate more effectively. Mid-market firms excel in this domain. Their leaders tend to remain close to customers. Their teams understand operational pain points firsthand. Their processes can pivot quickly in response to customer needs. AI enhances this closeness. It allows firms to personalize interactions, anticipate behaviors, and tailor services with precision. Enterprises, by contrast, often design solutions around internal structures rather than the customer's actual job. Their scale introduces rigidity that makes rapid customer-centric innovation difficult. AI amplifies this rigidity, creating a competitive opening that mid-market companies are uniquely equipped to exploit.

These theories illuminate why mid-market firms begin the AI era with structural advantages. But when we examine real-world operations, the advantages become even more visible. Consider a mid-sized manufacturing firm that implemented AI-driven predictive maintenance across its production lines. Previously, the firm relied on scheduled inspections and manual monitoring to prevent failures, a system prone to inefficiency, downtime, and unplanned repair costs. AI changed the equation. By analyzing real-time sensor data—vibration, temperature, acoustic signatures—the company could predict equipment failures before they occurred. Downtime declined by nearly half, production throughput rose, and overall equipment effectiveness improved significantly. None of these gains required expanding staff or investing in new physical infrastructure. The company simply infused intelligence into its operations, achieving results previously limited to global manufacturers with sophisticated analytics divisions.

A different example emerges from a genomics company struggling with the complexity of its data ecosystem. Scientists faced a bottleneck: While they possessed the expertise to interpret genomic data, the process of querying, correlating, and analyzing datasets was slow and cumbersome. By integrating an AI-powered natural language interface capable of interpreting complex queries, generating insights, and visualizing patterns instantly, the company transformed its research cycle. Scientists could ask questions conversationally—"What anomalies exist in this dataset?" or "Show me correlations between this expression pattern and known variants"—and receive immediate answers. The result was not a replacement of human expertise but an amplification of it. Projects moved faster. Insights deepened. The firm's scientific capacity expanded without increasing headcount. For a mid-market research organization, this was the kind of performance leap once thought impossible without enterprise-level resources.

These examples illustrate a broader truth: Operational AI equalizes not by diminishing excellence, but by democratizing it. It allows mid-market companies to perform at a level once accessible only to giants, while preserving their cultural agility, intimacy, and speed. AI does not merely compensate for what mid-market firms lack. It enhances what they uniquely possess.

This dynamic becomes even more pronounced when considering the constraints facing large enterprises. Fortune 500 companies carry the weight of legacy systems, entrenched hierarchies, risk-averse cultures, and incentive structures tied to old ways of working. Implementing AI requires re-architecting workflows, shifting power dynamics, and challenging deeply rooted norms. Mid-market organizations can bypass this struggle. They are not burdened by decades of accumulated process debt. They can redesign their operations more holistically, experiment more freely, and scale successful initiatives rapidly. In an era defined by speed and learning curves, this agility is decisive.

In short, AI rewrites the economics of the firm in ways that systematically advantage mid-market companies. And those advantages grow stronger as adoption accelerates.

The Economic and Societal Upside

While the transformation inside firms is profound, the broader economic implications are equally significant. Public discourse often frames AI as a threat to employment, conjuring images of widespread job displacement. Yet history suggests the opposite. Every major technological revolution initially disrupts tasks but ultimately creates new roles, new industries, and higher overall employment. Mechanization reduced the need for farm labor but unleashed manufacturing growth. Computing reduced clerical work but created entire sectors. The internet automated information exchange but generated millions of new jobs in services, logistics, e-commerce, and digital infrastructure.

AI follows this same lineage.

AI is fundamentally a labor-augmenting technology. It automates repetitive tasks, accelerates information processing, and enhances decision quality, but it does not eliminate the human capacity for creativity, empathy, problem-solving, or judgment. Instead, it elevates human work by removing the friction that consumes time and energy. Customer service representatives spend less time searching for answers and more time building relationships. Analysts spend less time cleaning data and more time interpreting it. Engineers spend less time diagnosing issues and more time innovating.

The productivity expansion unleashed by AI increases output, which increases demand, which increases employment. Organizations that previously lacked capacity to expand now find they can grow without proportionate cost increases. Mid-market firms, in particular, benefit from this effect. They frequently face labor shortages and operational bottlenecks that limit growth potential. AI lifts these constraints, enabling expansion into new markets, new product lines, and new customer segments. As they grow, they hire more people, both to support expanded operations and to fill new roles created by AI, such as AI operations managers, data stewards, workflow designers, and customer experience specialists.

From a macroeconomic perspective, the effect is substantial. According to McKinsey, AI could add trillions of dollars in global productivity gains in the coming decade. Goldman Sachs projects as much as a 7% increase in global GDP as AI diffuses across the economy. The Bureau of Labor Statistics forecasts rising demand for analysts, technologists, and human-centered roles as AI reshapes work. These gains do not originate solely from enterprise adoption. They come from broad-based transformation, particularly among mid-market firms that collectively generate a large share of economic output.

Furthermore, AI expands the frontier of what mid-market firms can offer. Predictive insights, personalized services, real-time analytics, and AI-driven product enhancements open new business models. Each model creates new kinds of work. Entire professional categories will grow because AI increases the rate of innovation and the speed at which companies can bring ideas to market. AI also elevates workforce skills by shifting labor from routine tasks to higher-value contributions, increasing wages in sectors where productivity gains translate into greater profitability.

The fear that AI will eliminate work misunderstands the nature of economic growth. AI eliminates tasks, not jobs. It transforms roles, reshapes workflows, and increases the value of human contribution. Economies grow not when work is preserved, but when output expands. Mid-market AI adoption will be a powerful catalyst for this expansion, both by increasing firm-level productivity and by unlocking new employment opportunities. It represents not a threat to workers but a pathway to national economic renewal.

Conclusion

The rise of operational AI marks the most significant shift in competitive dynamics since the advent of computing. For decades, mid-market firms operated under the shadow of enterprise incumbents whose scale granted them advantages in efficiency, specialization, and capital. Those advantages are now eroding. AI compresses coordination costs, automates the routines that once required large teams, and disseminates intelligence across the organization. Speed of execution, closeness to customers, clarity of purpose, and willingness to redesign workflows—traits inherent to many mid-market firms—now matter far more than headcount or geographic reach.

This is the strategic moment mid-market leaders have been waiting for. But like all great opportunities, it comes with urgency. AI rewards early action. Firms that operationalize AI today will harvest compounding benefits—more data, faster learning, smarter workflows, superior customer experiences—that late adopters cannot easily replicate. Waiting is not a neutral decision. It is a decision to fall behind.

The companies that seize this moment will not merely improve efficiency. They will redefine what their industries expect. They will offer customers speed, intelligence, and personalization that incumbents cannot match. They will attract talent seeking modern tools and innovative environments. They will raise the competitive bar in ways that force slower organizations into defensive positions. They will become the new leaders.

But leadership requires resolve. It requires CEOs, boards, and private equity partners to treat AI not as an IT initiative but as a strategic transformation. It requires investment in data, processes, training, and new forms of decision-making. It requires courage to reimagine how work gets done. And above all, it requires moving now before the advantages of early adoption give way to the penalties of falling behind.

Mid-market companies have never been better positioned to lead. The question is no longer whether AI will reshape your industry. It will. The question is whether you will reshape your organization quickly enough to define the next era of competition. The firms that do will not just compete with the giants. They will become the giants.

About the Author

Alejandro Laplana is the founder and CEO of technology company Shokworks Inc., where he has developed over 500 digital platforms and apps for some of the world's most exciting brands and companies, including FC Barcelona, La Liga, Real Madrid, P&G, Dupont Chemicals, the Las Vegas Raiders, the Tampa Bay Buccaneers, the

Pittsburgh Steelers, Fox Sports, HBO, the United Nations, and the Government of the UK.

Shokworks holds a staff of 500-plus full-time employees that design, develop, and operate highly scalable software solutions for large brands, operators, and well-capitalized start-ups. Laplana has generated over $1 billion for Shokworks and its partners in revenue, cost- savings, and direct investment. Shokworks and its partners have amassed 10 million users in aggregate, tokenized over $20 million in assets and have been named Apple App Store's "App of the Day" twice. Alejandro is also the founder and managing partner of VC Fund, Kevlar that bootstrapped its way to a portfolio of 30 companies, with a total portfolio value of $100 million. Laplana has personally invested in over 100 start-ups and sits in multiple boards of directors and advisors. Laplana was voted "Entrepreneur and Investor of the Year" by Bank of America and the DEC Network in 2022.

In addition to Shokworks and Kevlar, Laplana serves on Start-Up Chile's board of judges for upcoming startups and serves as a mentor to Ripple Labs and Techstars's Accelerator. Laplana has been invited to speak at forums hosted by J.P. Morgan, Cantor Fitzgerald, Bank of America, The DEC Networks, Techstars, Start-Up Chile, the Colombian Congress, and the United Nations. Laplana appears and contributes frequently on Bloomberg, Nasdaq, Rolling Stone, NBC, Cointelegraph, and Forbes. Laplana holds a bachelor's degree from Syracuse University and a master's in management of technology from New York University.

Email: alaplana@shokworks.io

CHAPTER 19

HUMANITY AT THE CENTER OF AI & PUBLIC SERVICE

By Brandon W. Ledford
AI & Human Capital Transformation
Pittsburgh, Pennsylvania

It has become appallingly obvious that our technology has exceeded our humanity.

—Albert Einstein

Artificial intelligence has arrived with breathtaking speed. In just a few short years, AI systems went from something fun that we used to make fast grocery lists or poems similar to Shakespeare's to tools that are used to create deep fakes, assist military operations, and increasingly, replace workers regardless of industry. For some, this is a miracle. For others, it's a nightmare. And for the millions of public servants who keep our society running, it brings uncertainty at best and hopelessness at worst.

The questions being whispered in city halls and municipal corridors are real. "Will artificial intelligence and robots make my career obsolete?" "Can an algorithm really replicate human judgment?" "Will the institutions we rely on still have a soul?"

These are reasonable questions; however, they are also incomplete.

I think that the most important question of the AI era is not what machines will do, but who we choose to be as humans while we build them.

I'm hopeful that we can build a world in which AI does not have to hollow out our workforce, erode trust, and widen inequality. In fact, if used wisely, it can do the opposite. Humanity-based AI can elevate work, restore dignity, and help the people who serve our communities do their jobs better, while we all live fuller lives in the process. That future is not guaranteed, but it is achievable.

The Revolution Is Not Technical. It's Human

Much of today's AI conversation focuses on technical breakthroughs: larger models, faster processing, more automation. These advances obviously matter, but they are not the real revolution. The real shift is human.

AI is already changing how decisions are made, how work is organized, and how value is measured. It's challenging long-held assumptions about expertise, productivity, and even identity. The disruption will not be evenly distributed. It will be felt most acutely by workers navigating outdated systems, limited resources, and rising expectations, and this is particularly true in state and local government.

As a White House intern in Washington, DC, a government consulting executive, and an elected official, I have spent much of my career working alongside public servants: the permit reviewer trying to keep a city moving, the caseworker managing impossible caseloads, the HR director holding together a workforce stretched thin, the borough manager struggling to provide services with rising costs and a diminishing tax base. These roles are deeply human and rely on judgment, empathy, context, and trust. AI should be used to strengthen those qualities, not replace them.

When AI is framed purely as an efficiency tool, people, and everything about them, from their loved ones to their mortgages to their hopes and dreams, become secondary. But when it is designed as a partner, something powerful happens. Work becomes easier and more meaningful, institutions become more responsive, and communities become stronger.

A Survival Story: Why I Put Humanity First

My perspective on technology is shaped by more than policy or professional experience. It is shaped by survival.

When I was nine, I was diagnosed with the African form of Burkitt's lymphoma. Burkitt's lymphoma is rare, but the African form is practically unheard of. Doctors found a treatment program from an oncologist in France, and I was given a 5% chance of survival. Fortunately, I had one of the best medical teams in the world from the Children's Hospital of Pittsburgh, and after a tough five-year battle using every experimental treatment and chemotherapy and technology available, I was blessed to be in remission.

Upon returning home, my mom was let go of her job at Altoona Hospital for missing time being with me. This was in 1994, right as the Family and Medical Leave Act (FMLA) was being passed, and a kind pro bono attorney took our case. Since my mom was out of a job and the medical bills were piling up, we were on public assistance, and I received free lunches at school. After witnessing many experiences of seeing the government protecting folks in their times of greatest need, I knew I wanted to dedicate my life to that cause.

I survived. I became the first person in the US to beat that specific cancer. That experience stripped away the noise and left me with a permanent understanding of what matters: faith, family, purpose, and the fragility of time. It's why I've dedicated my life to public service. And it's why I refuse to let technology overwrite the human pulse of our communities.

Local Government: The Front Line of the AI Era

While global technology firms dominate headlines, the real AI story will be written locally.

Cities, counties, school districts, and public agencies are where AI meets real life. They are where policies become services and technology touches people's daily experiences. They are also where trust is either built or broken.

In my work with local governments, I have seen AI dramatically improve service delivery: more access to incredible talent, faster permitting, smarter workforce planning, more responsive human services. I have also seen the risks when tools are deployed without guardrails and when speed outpaces understanding and innovation outpaces trust.

Local governments are uniquely positioned to model responsible AI use. They are closest to the people they serve and held to the highest standards of accountability. When they get AI right, they show what ethical innovation looks like in practice.

The CIVIC Framework: A Local Government Blueprint for Human-Centered AI

If AI is going to reshape government, it must strengthen the very thing government exists to protect: the civic fabric.

That's why I propose the CIVIC framework, which is a model designed specifically for local government leaders navigating AI transformation. I propose this because AI in Washington looks different than AI in a borough office, and AI in Silicon Valley looks very different than AI in a school district.

Local government is civic infrastructure. So our AI strategy should be, too.

C—Community Impact First

Before efficiency, cost savings, and automation, ask, "How does this improve life for residents?"

Every AI use case in local government should answer three questions:

1. Does it improve service delivery?
2. Does it increase fairness?
3. Does it strengthen trust?

If the answer is unclear, pause.

AI should not exist to modernize government for its own sake. It should exist to improve lived experience.

I—Institutional Accountability

Local governments are stewards of public trust. That means AI systems must be:

- Auditable
- Transparent
- Governed by policy
- Subject to human review

There shouldn't be black boxes deciding zoning, benefits, hiring, or enforcement. If a resident asks, "How was this decision made?" there must be an answer. Accountability is not a constraint on innovation, but rather it is what makes innovation legitimate.

V—Value the Workforce

AI, as well as decision makers using AI, must not treat public servants as expendable. AI in local government should:

- Remove administrative burden
- Expand capacity
- Enhance expertise
- Support reskilling

The goal is not to shrink government blindly. The goal is to make the people inside government more effective, more empowered, and more focused on meaningful work. If AI adoption creates fear instead of empowerment, implementation has failed.

I—Inclusive Access

Local government serves everyone, so AI must not widen the digital divide. That means:

- Designing for accessibility
- Supporting residents without technical fluency
- Preserving non-digital service channels
- Monitoring bias in models and outputs

Efficiency for some cannot come at the exclusion of others. Inclusion is not optional in public service.

C—Civic Trust by Design

Civic trust is local. When residents interact with government, they are not thinking about algorithms. They are thinking:

- "Was I heard?"
- "Was I treated fairly?"
- "Can I trust this system?"

AI must be implemented in ways that increase legitimacy, so public sector leaders must get good at:

- Transparency reports
- Clear AI use policies
- Public communication
- Community engagement before deployment

The CIVIC Framework does three things:

1. It anchors AI to the mission of local government.
2. It centers people, both residents and workers alike.
3. It reinforces that technology is a civic choice, not just a technical one.

It's simple enough for anyone in the workforce to remember and concrete enough for a city manager to implement. Even more importantly, it's values-driven enough to withstand political shifts.

How Governments Can Thrive in the AI Age

AI's disruption is real, but it is not something to fear. Public sector leaders and institutions who approach this moment with curiosity, adaptability, and values-driven leadership will not just survive. They will lead. Here is how:

1. Learn to work with AI, not against it.

Treat AI as a junior partner. Fast, capable, but in need of guidance. Use it to amplify your strengths.

2. Build AI literacy.

You do not need to code. You do need to understand how AI works, where it fails, how to evaluate its outputs, and how AI can help you do your job of improving communities better.

3. Reclaim time for what matters.

Use AI to automate the noise, so you can focus on judgment, relationships, and impact.

4. Invest in lifelong learning.

Careers will be nonlinear. Curiosity and adaptability will matter more than any single credential.

5. Anchor the work in values.

Ethics, fairness, and community impact are not soft skills. They are leadership skills.

6. Create bold AI use policies.

While AI technology rapidly changes, what matters to our people and our communities does not. Governments should still lead in AI policy development. My community (the Borough of Dormont, Pittsburgh, PA) was able to pass an AI Use Policy that clearly stated that AI will not be used to replace jobs or create art that can be done by our vibrant artistic community. What matters the most to your community and how can you help to push government forward with creative, human-inspired policy?

A Future Worth Building

AI's future is not predetermined. It will reflect the choices we make about who benefits, who bears risk, and what we value. If we lead with fear, we will get division. If we lead with greed, we will get inequality. If we lead with humanity, we can get something better: A future where technology serves people. Where work has dignity. Where innovation strengthens communities rather than hollowing them out. That future is possible. But it requires leadership at every level. Not just from technologists or executives, but from teachers, city managers, case workers, mayors, and public servants who show up every day to make communities work. They are not standing in the way of progress. They are progress. With AI designed around the human heart, they can help lead us into a future that is not just smarter but wiser.

About the Author

Brandon Ledford is a senior vice president at MGT (mgt.us), a national public-sector consulting firm, where he leads work at the intersection of human capital, artificial intelligence, and government transformation. He has advised state and local governments across the United States on workforce strategy, operational modernization, and responsible AI adoption. A former US Supreme Court and White House intern, Brandon holds a Master of Public Policy and is pursuing a master's in artificial intelligence management at Georgetown University. A cancer survivor and local government leader, he advocates for ethical, human-centered innovation that strengthens communities. He lives in Pittsburgh with his wife, Kate, and two daughters, Charlotte and Eliza.

Email: bledford@mgt.us

Website: www.mgt.us

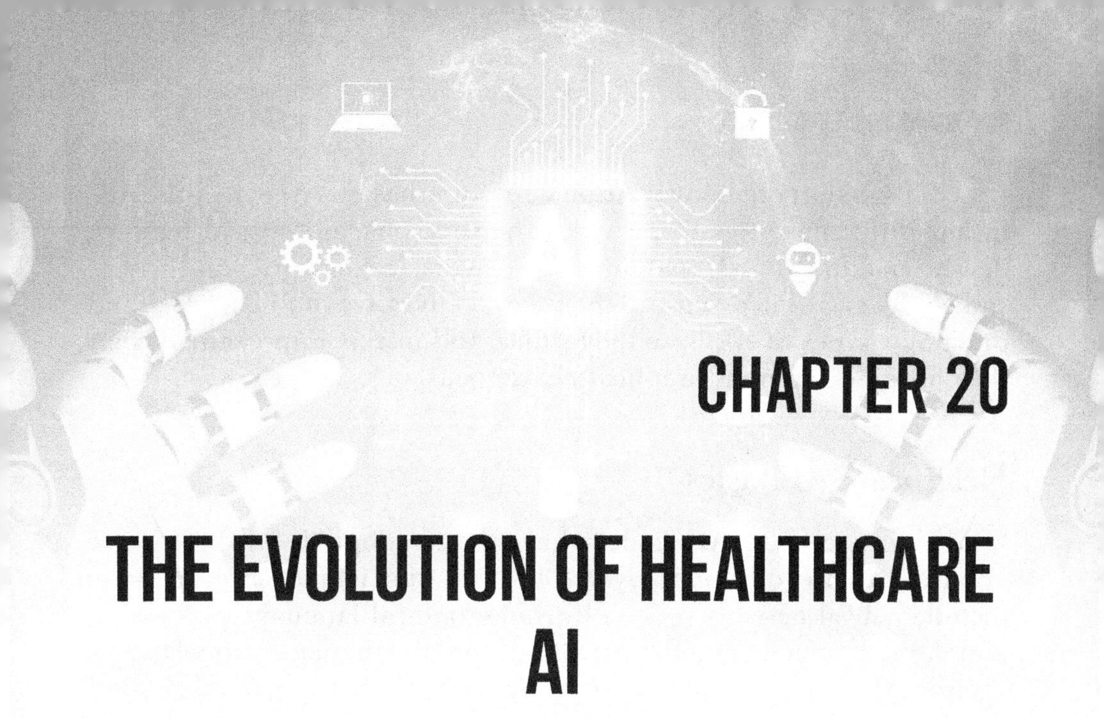

CHAPTER 20

THE EVOLUTION OF HEALTHCARE AI

By Ron Ozminkowski, PhD
Founder, Analytic Strategies & Consulting, LLC
Ann Arbor, Michigan

Let's begin again, begin the begin
—REM, from "Life's Rich Pageant"

The healthcare AI anthology began anew when AI developers turned their efforts to saving and improving human lives at less cost, with better technology. They want AI to hit healthcare on all cylinders. Their visions are indeed praiseworthy, yet scientific, technical, political, and regulatory hurdles make healthcare perennially slower than other industries to adopt advances in AI.[1]

How far, then, can the healthcare train go? So far, the lesson is clear: AI's potential can be harnessed only if developers successfully navigate the technical, business, political, and economic challenges every innovation faces when trying to scale in this challenging industry.

Let's start our conversation with some basics. "AI in healthcare" means different things to providers, payers, pharmaceutical makers, device manufacturers, patients, caregivers, regulators, and other stakeholders. To make sense of their perspectives, we should understand the major types of AI, how they differ, and just as importantly what they have in common for a healthcare focus.

Healthcare AI Basics

Several interrelated AI models have been adopted for healthcare applications. In addition to over 100 forms of machine learning, these include neural networks, deep learning, natural language processing and large language models (LLMs), small language models, and agentic AI. Thus, the notion of healthcare AI correctly goes beyond chatbot LLMs, such as ChatGPT, that are at the top of many minds today.

Notably, every form of AI makes heavy use of machine learning (ML). Indeed, AI cannot exist without this basic building block. ML works by using computers to find patterns in data. Those data might include numbers, biomarkers, gene expressions, words, phrases, pictures, X-rays, or scans, and these are often carved into parts called tokens. Statistical and mathematical processes on tokens are then carried out to predict one or more outcomes of interest to healthcare stakeholders.

Such outcomes might include the probability of developing a disease, being hospitalized to treat it, the probability that a mass or polyp detected in the body is benign or cancerous, the likelihood that a particular drug or procedure will help fight the disease, the likelihood that insurance will cover treatment for it, or the probability that someone will be seriously disabled or die from it. These are just a few of the thousands of questions that can be addressed by ML and related forms of AI.

Early AI in Healthcare

AI in healthcare is the application of any of the model types mentioned above to address health-related questions. Early applications were dominated by ML via advanced statistics. These often involved "supervised analyses" of events and suspected event predictors which could be found in data and labeled as such in advance of the analysis.

Supervised analyses often used multiple regression or logistic regression techniques, naïve Bayes or Bayesian regression techniques, support vector machines, decision trees, random forests, early forms of neural network analyses, and other techniques to predict health-related outcomes.

Many more supervised techniques have developed over time. So have "unsupervised analyses," which focus on finding outcomes that cannot be labelled in advance. Over 100 descriptions of supervised and unsupervised ML models can be found at this URL: https://medium.com/data-science/what-do-101-dalmations-and-machine-learning-have-in-common-9e389b899df3; all have already been applied to healthcare examples.

Even So, a Lot Was Missing

Thousands of clinical, economic, and business insights have been created by ML over the last 50 years. Many arose from using health insurance enrollment and healthcare utilization data coded numerically and contained in large linkable tables. Until recently, analyses of tabular data could not leverage information in images, written text, insights described in electronic medical records (EMRs), or conversational speech, even though these account for over 80% of all healthcare data.[2]

The AI world changed around 2010 though. Breakthroughs in natural language processing (NLP) allowed healthcare stakeholders to search written, spoken, and image data at near light speed in real time, trying to generate the best answer to their queries. Very quickly thereafter, AI processes for tumor recognition were built, as were sharable summaries of EMR text and images. Thus, AI evolved to address many more queries from clinical specialists.

As NLP and image processing advanced even further, many developers took advantage of deep learning models mentioned above. Over the next dozen years, deep learning became the building block of much better NLP and imaging techniques. These then begat large language model (LLM)-based products such as ChatGPT, which was released in 2022.

Now ChatGPT and its competitors are used frequently by hundreds of millions of people around the world to summarize text, write code for computer programs, create chatbots and AI agents, and write stories. Other tools such as Meta AI and Midjourney can produce real-looking images and short films. These technologies can greatly enhance doctor/patient/pharma/insurance relationships, if handled properly.

What If vs. What Is

Imagine that patients and their advocates, providers, representatives from insurance companies, drug and device companies, employers, regulators, and policy wonks would get together to discuss how to develop, provide, and fund healthcare services in an ethical, empathetic way. Their efforts would generate optimal patient outcomes at reasonable costs and with sustainable profits. Everyone would win.

Good luck with that!

Instead, the US has an unorganized system where all parties say they put patient interests first, but often don't.[3] For example, many begin by using AI to maximize their own profits. Eventually, at least some of their interests may overlap with patient interests, but controversies frequently arise when financial interests collide.

Provider and Business Perspectives

One bone of contention between providers and administrators is whether AI should be used to replace medical or support staff, because AI can do large chunks of their work much faster. Nobel Prize-winning physicist Geoffrey Hinton, a pioneer in AI, says the billions of dollars invested in AI in healthcare and other industries cannot be recouped unless AI replaces a substantial proportion of workers.[4]

For example, radiologists look for anomalies in X-rays, CAT scans, MRIs, and PET scans, jobs which AI can do just as well and much quicker. The same goes for many jobs conducted in medical laboratories and pathology work. AI can do a lot of work otherwise completed by physical and occupational therapists too, among others.

Whether AI eventually replaces many of these professionals will depend on whether it learns to replicate advanced human thought processes accurately and safely.[5] It has a long way to go here because thinking and reasoning are not AI strong suits. No machine understands how humans think so it has been quite difficult to mimic thinking and reasoning when developing AI.

What Doctors Want from AI and Where They See It Going

On a recent podcast two leaders of the Journal of the American Medical Association's AI team discussed several ways they would like to see AI develop in 2026 and beyond.[6] These included:

1. Whether and how patients should be informed when their providers use AI
2. Randomized comparisons of the efficacy of competing AI applications
3. Translating ChatGPT or other LLM output into many languages, in real time
4. Translating oral conversations in real time, so providers can communicate better with patients whose first language is not the same as their providers'

5. Reliable and valid evidence that using AI will reduce provider burnout
6. Better applications to prevent life-threatening infections among hospitalized patients.

In the next few years, AI is likely to go well beyond its simple search and chatbot uses.[7] It will focus more on patient history taking, helping doctors diagnose patient conditions more accurately. AI will also make better use of digital scribes that document doctor-patient discussions. Whether and how well these work at scale remain to be seen.

What Nurses and Allied Health Professionals Want

To maximize its potential, AI should be baked into every provider's workflow. It must work directly with, or ideally be integrated into, all the EMR systems providers use. Most patient visits are short, lasting just a few minutes, and time is of the essence. AI products requiring providers to move from one unintegrated system to another waste too much time, failing to benefit providers or patients much. This is one reason why AI has not caught on as rapidly in healthcare as in other industries.

At the other end of the spectrum, psychologists, physical therapists, and occupational therapists are examples of providers who spend more time with patients than medical doctors do. Psychology, PT, and OT visits often last an hour or so. Much of that time is spent diagnosing problems, describing treatment options, offering treatment and treatment adjustments, and documenting patient progress. LLM-generated patient histories and digital summaries can help these providers immensely.

Nurse practitioners, registered nurses, or licensed practical nurses can also benefit from AI. Their patient visits tend to be short yet recur frequently. These present challenges to keep up with real-time developments, especially across staff shift changes. AI-generated chart reviews and recommendations may help them too.

How Do Payers Use AI?

Payers include self- and fully-insured employers, commercial insurance companies, and the government (via, for example, Medicare, Medicaid, TRICARE, CHAMPVA, and many state disability programs). Just a few examples of how they use AI are noted here.

Self-insured employers pay administrative-services-only (ASO) fees to insurance companies to do what insurance companies do best—move money. ASO fees cover the costs of setting up networks of medical providers. ASO fees also cover the costs of designing insurance options that employers offer to workers and their families. The ASO fee covers the costs of annual insurance enrollment processes too. ASO fees also cover the costs of estimating premium and out-of-pocket payment levels for healthcare provided within and outside provider networks.

AI by or for all payers can help with these processes. For example, Optum and other insurance partners provide AI services to process insurance claims and handle other internal payer operations. Optum and others also provide well-being and chronic care management programs which use AI to help improve the quality of care or reduce medical expenditures.

What Patients Need from AI

Patients often say they want better access to care, lower costs, better insurance coverage, higher-quality care, and much less frustration navigating the medical system. Can AI help? Consider the following example reported in the New York Times.[8]

Wendy Goldberg, a 79-year-old from Los Angeles, emailed her doctor for some nutritional advice. Unfortunately, his response made it clear he wasn't really listening, and his reply was irrelevant. Ms. Goldberg turned instead to ChatGPT, which provided a detailed daily nutrition plan, exactly what she requested.

The NYT reporters said many providers see this kind of doctor/chatbot/patient relationship as intrusive, alarming, or economically threatening. Others are OK with it if the chatbot does not hallucinate inappropriate or dangerous responses. AI did not recommend anything contraindicated for Ms. Goldberg, and her story ended well.

In other instances, though, AI may disregard or misinterpret patient circumstances. Horrifying consequences can arise if the provider cannot intervene in time to rebut the AI. Several illustrations of this have been described among patients with vexing or life-threatening mental health problems because AI downplayed or ignored their symptoms or recommended dangerous options for the patient to consider.

The lesson is clear for providers and patients alike: Chatbot users should proceed with caution but are not likely to ignore the new AI reality. Providers are stretched for time and cannot be ever present, even by email or internet. Providers, patients, and caregivers must figure out how best to use AI, whether it be in chatbots; AI agents; electronic gadgets like heart rate, blood pressure, or blood sugar monitors; or other innovations.

The Prior Authorization Problem

Prior authorization means providers must get permission from the insurance company to get paid for the treatments they want to prescribe or apply. Without AI, prior authorization can easily take days, weeks, or even months as humans search and summarize relevant literature to base their decisions on.[9] Such delays often complicate patient treatment and can reduce quality of care. Delays also increase out-of-pocket costs if insurance payments are ultimately denied and the patient must foot the bill.

Incorporating AI correctly, however, can avoid these frustrations. Unfortunately, patients and providers argue that AI-based prior authorization is frequently used to the opposite effect, denying insurance payments for many necessary treatments. Appealing the AI findings can then lead to the same unnecessary or dangerous delays in treatment that AI was meant to prevent. In some cases, those delays can be life-threatening.

On the other hand, insurers argue that AI-based treatment authorizations are typically correct more often than human authorizations are and may even prevent many harmful or clearly inappropriate treatments.

There is no easy solution to the prior authorization battle; that controversy has waged for decades now. What is clear is that AI should not be used to fix this problem by itself. Humans still need to guide its appropriate use. Providers, insurers, and others on different sides of the prior authorization issue must, therefore, negotiate to figure out how best to apply AI for medical treatment.

Summary

AI in healthcare has been around for over 50 years, starting with heavy duty ML to learn which factors have the most influence on health, healthcare, and health expenditures. More recently, ChatGPT and its ilk have greatly expanded the scope of AI in medicine. They allowed widespread use of AI by patients and their caregivers. They also facilitated or enhanced the work of insurers, providers, medical devices, drug developers, and researchers.

AI can surely increase the speed of medical innovation, but it can also increase the short-run costs of care until patents for new discoveries expire and economic competition forces prices for AI-enabled care to fall. Multimillion-dollar price tags for new cancer drugs and genetic therapies are examples of things that may break the bank if current economic conditions prevail. Whether AI will really help usher in an age of less costly and more effective care remains to be seen.

This chapter provides a basic understanding of how AI works in healthcare and what its potential for good or ill may be. Many more issues remain to be explored though, so I leave the reader with some thoughts about what else is or should be coming.

First, as mentioned earlier, small language models (SLMs) are one of many AI model types. SLMs are a brand new development, as is another AI model type called a method-of-expert (MoE) approach.

Both can reduce the resource costs of using AI substantially. SLMs use a healthcare organization's own, or similar, small datasets as the most relevant inputs to its AI queries. MoEs, by contrast, select hundreds or more large chunks from huge AI sources and run analyses on each one simultaneously before combining results to find the best answer to a query. Go to https://kimi-k2.com/ for more information about how MoEs work and to try one out.

What Else?

The healthcare AI ocean is vast, and more issues must be settled. Sigrid Berge van Rooijen addresses some of these in several LinkedIn posts that came across my feed in late November/early December, 2025. [10] These include:

1. Ever-present data and personal security risks
2. Poor descriptions and illustrations of how and why AI generates answers, leading providers to question its output
3. Biases in AI arising even when hallucinations are not present
4. The need to establish a greater sense of ethics and accountability for developing and using "responsible AI"
5. Whether AI utilization in healthcare should be licensed
6. Greater accuracy, reliability, and validity of AI-produced output

For more information about several of these topics, see my paper on how "garbage in yields garbage out" when analytics are carried out for all industries, including healthcare.[11] See also the National Academy of Medicine's guidelines for using AI.[12]

References

1. Haider J. Warraich and Dhruv S. Kazi, "Private Equity and the Future of US Health Care," *JAMA Health Forum* 6, no. 1 (January 2025): e244146, https://jamanetwork.com/journals/jama-health-forum/fullarticle/2841460.
2. Gwen, "Unstructured Data in 2025: Navigating the New Digital Frontier," *Design Tools Network*, March 15, 2025, https://designtoolsnetwork.com/unstructured-data-in-2025-navigating-the-new-digital-frontier/.
3. Jesse M. Ehrenfeld, "Artificial Intelligence in Medicine — Navigating the Transformation," *New England Journal of Medicine* 392 (January 2025): 101-103, https://www.nejm.org/doi/full/10.1056/NEJMp2513434.
4. "Godfather of AI Says Tech Giants are Using a 'Bad' Argument to Avoid Regulation," *Yahoo! Finance*, November 15, 2025, https://finance.yahoo.com/news/godfather-ai-says-tech-giants-184357806.html.
5. "Godfather of AI," *Yahoo! Finance*.
6. Michael S. Parmacek and Jonathan A. Epstein, "The Future of Academic Medicine in the Age of AI," *JAMA* 333, no. 2 (January 2025): 115-116, https://jamanetwork.com/journals/jama/article-abstract/2839717.
7. Lisa Rosenbaum, "The Future of Clinical Notes," *JAMA Internal Medicine* 185, no. 1 (January 2025): 12-13, https://jamanetwork.com/journals/jamainternalmedicine/article-abstract/2840192.
8. "Can an AI Chatbot Be Your Doctor?," *New York Times*, November 16, 2025, https://www.nytimes.com/2025/11/16/well/ai-chatbot-doctors-health-care-advice.html.
9. Gigi Yuen, "Why AI Must Be Clinically Trained to Transform the Prior Authorization Process," *Cohere*

Health, November 10, 2025, https://www.coherehealth.com/news/clinically-trained-ai-prior-authorization.

10. Sigrid Berge, "Activity," LinkedIn, accessed March 20, 2026, https://www.linkedin.com/in/sigridberge/recent-activity/all/.
11. Ron Ozminkowski, "Garbage In, Garbage Out," *Towards Data Science*, November 13, 2021, https://towardsdatascience.com/garbage-in-garbage-out-721b5b299bc1/.
12. Arjun K. Manrai and Isaac S. Kohane, "The Evolution of Medical Knowledge in the Era of AI," *NEJM AI* 2, no. 1 (January 2025), https://ai.nejm.org/doi/full/10.1056/AIp2500651.

About the Author

Dr. Ron Ozminkowski spent 15 years in VP and SVP/chief scientist roles, leading large groups of analysts. He has over 20 years of managerial experience, building groups from ground zero.

Ron is known for exceptional service to clients in The White House, federal and state agencies, life sciences companies, large insurers, hospitals, and health systems. He is also a prolific writer and speaker, co-authoring over 160 peer-reviewed and other publications and more than 160 professional conference presentations and speaking engagements.

Ron is frequently interviewed or quoted in the media and recently appeared on Stanford University' list of the top 2% most cited scientific authors in the world, from 1960 onward.

LinkedIn: https://www.linkedin.com/in/ronozminkowski/

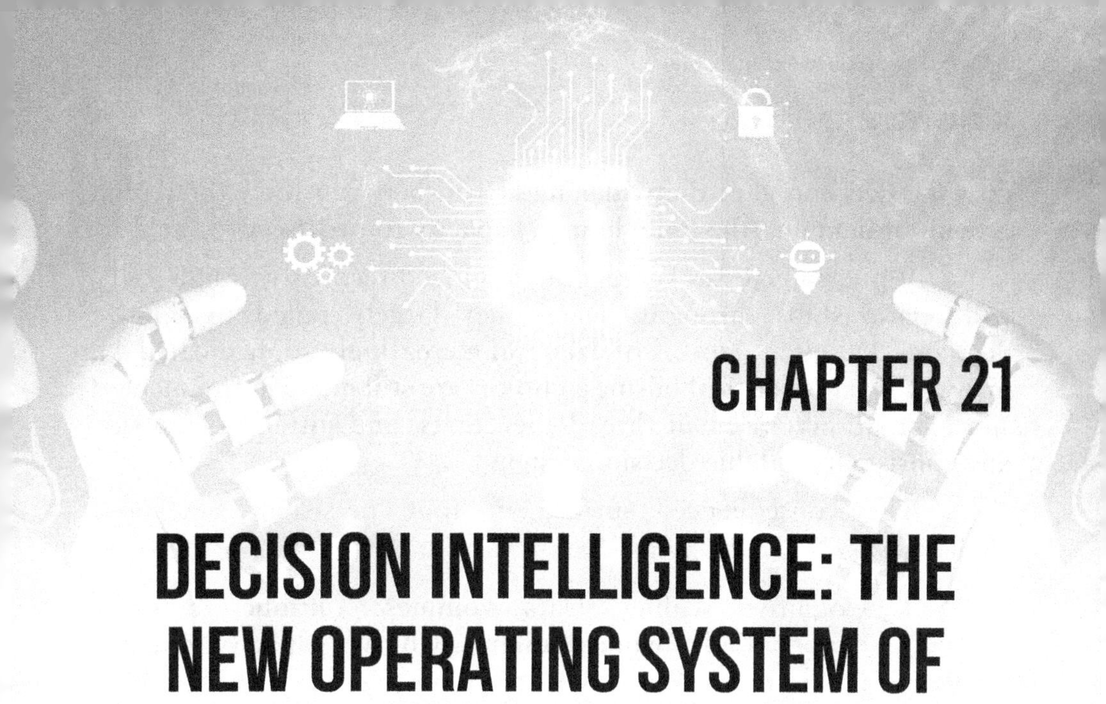

CHAPTER 21

DECISION INTELLIGENCE: THE NEW OPERATING SYSTEM OF ENTERPRISE GROWTH

By Sridhar Peddisetty
Decision Intelligence & Agentic AI Leader
Wake Forest, North Carolina

Action is the real measure of intelligence.
—Napoleon Hill

For years, enterprises equated growth with data, believing that more information would naturally lead to better decisions and stronger outcomes. Enterprises invested millions in ERPs, CRMs, business intelligence, and data platforms, assuming that visibility would automatically translate into better decisions and business value. Visibility didn't bring clarity; it created a level of complexity that outpaced human decision-making capability. Enterprises today are overwhelmed by rear-view-mirror reports and constrained by a growing decision gap: the widening chasm between the exponential growth of data and the finite human capacity to process it.

While enterprises successfully digitized records through ERPs and relationships through CRMs, they largely failed to digitize decisions. Despite terabytes of data and increasingly sophisticated AI, many of the most critical business choices are still made in fragmented silos, spread across email threads, meetings, and intuition—outside any consistent, scalable decision system.

The consequences surface in four persistent enterprise challenges:

1. Cognitive scaling: Data volumes continue to grow exponentially, while human cognitive capacity remains fixed.
2. Localized intelligence: AI adoption is fragmented and localized, delivering pockets of value rather than enterprise-wide impact.
3. Execution velocity: Market volatility demands speed and foresight that manual analysis and scenario planning cannot match.
4. Data fragmentation: Disconnected systems create blind spots that undermine forecasting and execution.

Over the past decade, enterprises have championed data-driven decision-making: building real-time dashboards, hiring data scientists, and producing ever more reports. Yet data remains fundamentally passive; it explains what happened yesterday, not what should happen next. As enterprises race to adopt AI, one critical element is often overlooked: how decisions are actually made and executed. AI's value is realized only when insights are translated into timely, informed action. This is where Decision Intelligence becomes essential for enterprises scaling growth.

Decision intelligence integrates data, AI, and business context with human reasoning to transform insights into decisions that are smarter, faster, and more consistent. It embeds learning, prediction, and action directly into everyday workflows, shifting enterprises from reactive analysis to proactive execution. Instead of diagnosing a bad quarter after the fact, decision intelligence simulates scenarios to prevent it. Each decision feeds a learning loop, making the next one with more informed intelligence. By digitizing expert logic and decision pathways, enterprises can scale human reasoning and execute growth strategies with discipline.

Decision Intelligence as the Operating System of Enterprise Growth

Decision intelligence (DI) serves as the new operating system (OS) for enterprise growth, replacing reactive reporting with proactive, predictive decision engines. Just as an OS manages resources, processes, and interactions, DI acts as the brain and nerve system of modern enterprises, which requires a shift from passive observation to active orchestration.

Within Decision Intelligence, there are three fundamental layers. From bottom up:

Foundation (Hardware): "What we observe" by ingesting data and signals from various source systems (ERP, CRM, legacy, etc.), and creating a unified "source of truth." Most enterprises suffer from "decision fragmentation" with sales data sitting in CRMs, financial data in ERPs, operational data in spreadsheets, and humans trying to bridge gaps manually. Foundation layer doesn't just integrate data. It creates the conditions for trust, speed, and intelligence at scale.

The foundational Data and Signal layer unifies fragmented data into a trusted core, breaking down data silos across ERP, CRM, legacy systems, and various groups to create a single, reliable foundation for decision-making. This layer ensures data is governed, contextualized, and ready for intelligence, not just stored. Enterprises have data platforms but still struggle to decide because of siloed data, varying definitions, missing context, low trust, and data showing up after decisions are needed. Unlike a traditional data platform, which is built for reporting, the Foundation layer is built for deciding by making data consistent, explainable, timely, and decision-aligned.

Brain (Kernel-Core): "How decisions are formed" is not by charts but by suggesting actions. Predictive analytics tell us what might happen but prediction without direction is just anxiety. This Decision Intelligence (DI) layer elevates us to prescription. While predictive analytics could tell it will likely rain, DI tells you to bring an umbrella. DI doesn't just flag a risk; it simulates outcomes and recommends the optimal path forward.

The Decision Intelligence layer helps you to better understand the interconnections within complex environments rather than analyze parts in isolation. This layer acts as the decision kernel that correlates signals across systems, applies AI and business logic, and surfaces what matters most. This layer accelerates action by transforming raw signals into prioritized, decision-ready insights. Paired with critical thinking, the ability to evaluate what-if scenarios, detect biases, and synthesize information, the Decision Intelligence layer enables an enterprise to navigate ambiguity and develop holistic, context-aware actionable decisions.

Execution (Applications): "What we decided and do" happens by enabling an enterprise to strategize growth. We still need human intelligence to handle the context (ethics/strategy/nuance) but with system-learning algorithms and feedback loops, DI applications keep getting smarter. Instead of a dashboard flashing red, the system presents a structured choice, simulating "Inventory is low. Reorder now to maintain service levels, or hold off to preserve cash flow?"

The Decision and Action layer embeds intelligence directly into decision-enabled applications and workflows, guiding human and agentic actions where decisions are made. This layer compounds value by turning insight into consistent execution, learning from every decision to continuously improve outcomes. It helps a business recognize how to form judgments, identify cognitive blind spots, and refine decision-making under uncertainty. Evidence-based reasoning, forming hypotheses, and validating outcomes can distinguish signal from noise and ground decisions in facts, not intuition.

DI as OS: Coming Together

From data foundation to intelligent decisions, to scalable execution, each layer plays a vital, interconnected role in helping enterprises continuously sense, learn, decide, and act with speed and precision. With DI as an OS, together, they unlock smarter, faster ways to deliver value across products, services, and the entire customer journey. A unified data foundation provides the clarity and trust required to see the enterprise as a whole. The decision kernel transforms this signal into understanding, applying AI and human judgment to anticipate what matters next. Intelligence is then woven directly into execution, guiding actions at the moment decisions are made. With every outcome feeding back into the system, enterprises don't just operate; they learn, adapt, and compound value, turning growth from a goal on paper into an enduring capability.

DI Bridges the "Last Mile" Between Insight and Action

Unlike a static dashboard, DI-enabled systems are constantly learning. Every decision made, whether accepted, rejected, or modified by a human, creates a feedback loop. DI systems use closed-loop learning, which means they continuously learn from past decisions and outcomes to improve future performance. This feedback mechanism ensures that decision models become more effective over time, driving ongoing margin improvement. Intelligence grows when enterprises learn from every decision they make.

If a recommended pricing strategy drives revenue up, the system reinforces that logic. If it underperforms, the weighting and assumptions are recalibrated. Over time, this creates a compounding asset: an enterprise that becomes smarter with every transaction, turning growth from a quarterly target into a continuous, automated capability.

DI as OS: Improving Effectiveness and Efficiencies

Decision intelligence when implemented as an enterprise operating system, fundamentally improves how effectively decisions are made and how efficiently the enterprise operates. By embedding intelligence into core workflows, DI transforms growth, cost, and risk management from reactive efforts into proactive, system-driven capabilities.

Here are some examples showing the benefits of DI:

Revenue Growth

- Opportunity prioritization using ML-driven scoring to focus effort on the highest-value deals
- Intelligent pricing and discount optimization that balances competitiveness with margin protection
- Client 360 intelligence to enable targeted cross-sell and upsell of products and services
- Pipeline-to-revenue acceleration by identifying and removing decision bottlenecks across the sales cycle

Cost Optimization

- Automated demand and workforce forecasting to align capacity with business needs
- Skills-based workforce planning through intelligent talent-to-demand matching
- AI-driven scheduling and capacity optimization to maximize utilization
- Reduction of non-billable or underutilized time through proactive deployment decisions

Risk Mitigation

- Early-warning signals for revenue churn, delivery risk, and margin leakage
- Scenario planning to prepare for market volatility and demand shocks
- Automated compliance, governance, and data-quality monitoring to reduce operational and regulatory risk
- Decision variance reduction by minimizing the "noise" of inconsistent human judgment by codifying best practices into a unified logic, ensuring the same high-quality outcome every time

When Decision Intelligence operates as the enterprise OS, effectiveness and efficiency are no longer trade-offs. Decisions improve, execution accelerates, and learning compounds, enabling enterprises that don't just react to change, but systematically outperform through it.

DI as OS: Business Impact

DI as OS doesn't just improve decisions. It builds an enterprise that learns faster, acts sooner, and compounds advantage over time.

Improved Decision Quality

- Contextual intelligence provides a 360-degree view of complex scenarios, neutralizing individual cognitive biases.
- With its predictive orchestration moving from historical tracking to scenario simulation, DI enables the forecasting of future trends and outcomes, allowing for proactive planning and resource allocation, where they will have the highest future impact.

- Real-time data analysis enables enterprises to make informed decisions quickly, staying ahead of the competition.

Greater Operational Efficiency

- The "zero-latency" workflow automates high-frequency micro-decisions, removing the "human middleware" that causes delays in supply chain, pricing, and staffing.
- When DI eliminates friction by integrating disparate systems (ERP, CRM, HRIS) at the logic layer, this leads to the elimination of data silos and manual re-entry, creating a seamless flow from signal to execution.
- DI enables continuous optimization with data-driven metrics acting as the system's "diagnostic tool," identifying bottlenecks and re-routing workflows automatically to maintain peak performance.

Accelerated Innovation and Adaptability

- Pattern discovery uncovers hidden correlations in customer behavior and market shifts that human analysts often miss, sparking new service offerings.
- With its agile experimentation using a "digital twin" of a business, DI enables low-risk A/B testing of strategies. You can simulate a new market entry or pricing model in the OS before spending a single dollar in the real world.

Sustained and Compounding Competitive Advantage

- With its velocity as a moat speeds the "decision cycle" in a volatile market, providing the ultimate advantage, DI enables you to respond to customer needs and market shifts while competitors are still waiting for their weekly reports.

- The learning flywheel feeds back into the system as every decision made is tracked and the OS becomes a compounding asset. Enterprise literally gets smarter with every transaction, making your competitive lead impossible to close.

Just as an OS manages the communication between hardware, kernel (core), and applications, DI serves as the connective tissue between fragmented data silos and strategic outcomes. It contextualizes data, automates low-level decisions, amplifies high-value human decisions, connects operational and strategic layers, reduces decision latency across the value chain, and creates consistency across sales, products, solutions, delivery, and operations (finance, people solutions, TA, and others) teams.

DI as OS becomes essential enterprise infrastructure: the center of gravity everything connects to. Always on. Always learning. Continuously optimizing. It is not a tool, a model, or a dashboard, but the system through which modern enterprises execute growth.

The Final Pivot: From Intelligence to Impact

True intelligence is not knowing more. It is deciding better. For over a decade, enterprises have been obsessed with "time to insight," spending millions of dollars in building data platforms and dashboards that promised real-time visibility. While this delivered unprecedented transparency, visibility alone does not drive growth. A dashboard can tell an enterprise what happened, but it cannot tell what to do next.

As we move into an era where every enterprise aspires to be AI-enabled, those that invest in DI will stand apart, not for how much data they have, but for how effectively they turn intelligence into action. The most intelligent enterprises are not the most informed, but the most decisive. The future-proofing enterprise requires moving beyond backward-looking questions and forward-looking decision-centric ones.

- The old way: "What is our current utilization?" or "How many hours are billable?"
- The new way: "Are we deploying the right skills at the right time?", "Which decisions are driving margin leakage?", or "Where will improving turnaround time immediately unlock revenue?"

Enterprise growth will no longer be powered by more tools or models, but by how intelligently decisions are orchestrated. Decision intelligence is the operating system for the modern enterprise. By combining subject matter expert logic, reducing decision variance, and automating foresight, DI transforms growth from a fortunate quarterly outcome into a disciplined, engineered capability. It is the engine that standardizes decisions, reducing variability, improving predictability, and scaling growth.

About the Author

Sridhar Peddisetty is a global technology leader specializing in Decision Intelligence and AI-driven transformation, focused on architecting autonomous decision systems that turn data into measurable business outcomes.

With leadership experience spanning Argentina, India, Japan, and the United States, he brings a strong cross-cultural perspective to building and scaling high-performing global teams. Grounded in an honors degree in Mathematics and a Master's in Computer Science, Sridhar combines deep technical expertise with pragmatic business execution.

He helps organizations evolve from traditional operations to Smart Ops by orchestrating data, AI, business context, and human judgment into intelligent workflows and decision-enabled applications. His work focuses on designing systems that not only generate insights but drive better decisions, enable execution, and continuously learn at scale.

Currently, Sridhar leads the Office of Decision Intelligence for a Fortune 500 enterprise, building AI-ready data platforms and scalable decision systems that improve efficiency, accelerate growth, and enable more consistent and accountable decision-making across the enterprise.

LinkedIn: https://www.linkedin.com/in/sridharpeddisetty/

CHAPTER 22

HUMANS & MACHINES TOGETHER: REDEFINING WORK IN THE AGE OF AI

By Stacy Proctor
Chief People Officer, AI Implementation Advisor
San Francisco, California

What makes us human, our deeply human skills will keep us relevant.

—Brené Brown

AI is here, and this is a leadership reckoning.

Humans working with machines is not new. When we transitioned to the knowledge-worker era, we resisted, then learned the advantages of how computing machines could improve quality and efficiency. But artificial intelligence (AI) seems different because of the threat that machines will replace our human intelligence. This is no longer a future-facing concept, a pilot project, or an abstract trend to track from a safe distance. Business boards and owners seeking easy productivity and competitive advantages are already shaping how decisions are made. Algorithms influence who gets hired, what automations are prioritized, which customers receive attention, and how success is defined. In many companies, AI is quietly embedded in workflows long before leaders articulate how it could impact their company's purpose and how the elimination of institutional knowledge has depleted their human competitive advantage.

This is why the moment we are in is not best described as a technological revolution. It is a leadership reckoning. The most important questions leaders face today are not technical. They are human. We focus on leaders because their AI experience will become the employee's AI experience.

Who do we want to be as humans while we build this future? Will human wisdom be recognized as the differentiator that improves outcomes for those we serve (customers, clients, patients, communities, and society at large)? How will we reckon with our culpability in the worst AI output if we don't continually and intentionally build outcomes that improve humanity in the world?

The organizations that thrive in the age of AI will not be the ones with the most advanced tools, the largest data sets, or the greatest number of automated processes. They will be the ones whose leaders earn trust, honor human dignity, and intentionally design how humans and machines with artificial intelligence work together. AI is powerful, and power without alignment to human purpose is dangerous. Speed without clarity erodes confidence. Automation without trust damages morale and performance.

There are numerous examples across industries in the United States where trust between employees and leaders was eroding before AI became widely visible to the general public, particularly in organizations that relied heavily on monitoring and control rather than autonomy and transparent communication.[1] For example, hybrid work became functionally necessary during the COVID19 pandemic as public health restrictions drove a large, rapid shift from onsite to homebased work in many officeeligible, nonessential sectors.[2]

"Nonessential" businesses and employees pivoted to remote or hybrid models at scale, especially in professional, technical, and business service roles that were feasible to perform from home.[2,3] Employees often stayed engaged, and when employees and leaders experienced adequate resources and support from each other under welldesigned remote and hybrid arrangements, such as clear expectations, sufficient technology, and supportive supervision, job satisfaction, engagement, and related indicators of trust and relationship quality tended to be maintained or improved.[3,4]

After years of sustained success driven by outcomes such as stable or enhanced productivity and lower turnover under remote and hybrid work, many organizations introduced topdown returntooffice (RTO) mandates, particularly daycount policies (for example, requiring employees to be in the office three days per week) that employees frequently perceived as arbitrary, disconnected from performance, muddling collaboration, and detached from role needs.[3,5] When these RTO mandates were poorly explained or weakly justified, they fueled skepticism about senior leadership's motives and reduced confidence that decisions were being made in employees' and customers' best interests.[5,6]

Instead of strategizing around the intentionality of being together in person for relationship building, complex problem solving, innovation sprints, and mentoring, many RTO policies were implemented as generic presence requirements and were associated with reductions in perceived fairness and autonomy, lower commitment, higher turnover risk, and the loss of key talent, particularly among high performers.[3,5,6] While there is strong evidence that remote and hybrid work often maintain or improve productivity at both the industry and firm level, there is also growing evidence that rigid RTO mandates damage engagement and increase attrition among valuable employees, which directly undermines organizational performance.[2,3,5,7]

Now, leaders have turned to machines with artificial intelligence to improve productivity. But AI will not rebuild trust amongst humans, and AI is dependent on human intelligence to succeed. Implemented without human wisdom, AI accelerates confusion, fear, disconnected processes, and disengagement at scale. Unfortunately, we have not recovered from the distrust built with the hybrid and remote work debate, so many employees are starting their AI journey with visible distrust. Therefore, the biggest risk in AI adoption is not the technology itself. It is the further erosion of trust, clarity, and humanity when leaders implement AI without aligning it with the organization's human purpose.

This is a defining moment. AI will not determine the success of future-of-work initiatives. Leadership will.

Humanity in an Age of Artificially Intelligent Machines

To understand why leadership matters so deeply in this transition, we must also acknowledge a simple truth: Humans are not just workers. People seek meaning in their contribution, dignity in their treatment, connection to others, and a sense that their efforts matter. Work has always been more than a transaction of labor for pay. It is a source of identity, belonging, and purpose. And those we serve with our products and services are people too, who seek human connection. AI challenges these deeply human needs when implemented solely to replace humans and treat those we serve as bots.

Cautious is the long-term competence we need as smart machines take on tasks that once defined roles. Many people experience a threat not just to their jobs, but to their sense of value. Questions surface quickly and quietly:

- If a machine can do this faster, what am I for?
- If decisions are automated, where does my judgment matter?
- If efficiency is the only goal, am I replaceable?
- If service is only provided by a bot, does anyone care about my experience?

These fears are not irrational. They are a natural human response to decades of eroding trust in institutions and workplaces, even as recent as the last couple of years with the hybrid and remote work debate. Many employees have already lived through waves of restructuring, cost-cutting, and change initiatives that promised progress but delivered uncertainty, disappointment, and setbacks. Promoting AI as a competitive advantage is not a neutral business decision but rather an affront to human spirit.

Humans cannot thrive, and organizations cannot succeed in the long term within systems where people feel devalued, invisible, or disposable. When employees feel replaceable, they disengage. When they disengage, performance declines. When performance declines, leaders seek more control, more automation, and more metrics, further eroding trust. This cycle is not new, but AI can accelerate it dramatically if leaders do not create and execute a strategy *with* employees to connect AI to improved human outcomes.

Fear should not be dismissed or minimized. It should be understood and addressed. Fear recedes when people are given transparency, clarity, and the opportunity to learn. Automating work that still allows employees to explicitly use human intelligence, their wisdom earned from their experiences, critical thinking skills, discernment, and conflict resolution matters, and it reduces fear. Creating a partnership between humans and machines is the new essential leadership skill.

This requires a shift in how organizations think about capability. Employees must be supported not only to learn how to use AI tools, but to articulate and strengthen the human intelligence that machines cannot replicate, including ethical reasoning, understanding emotions, creativity, empathy, and relationship-building. These are not soft skills. They are essential skills for leading and working alongside machines.

People are not resisting change. They are resisting being devalued.

Humans and Machines: Redefining the Meaning of Work

Much of the public conversation about AI is framed as a competition, as business decisions have focused on humans versus machines. This framing is not only misleading but also counterproductive. Machines excel at processing vast amounts of data, identifying patterns, and executing repetitive tasks with speed and consistency. Humans excel at context, nuance, values-based decisions, and human connection. The future of work depends on leaders who can design systems where all these strengths matter, are complementary, and are channeled with purpose to improve outcomes for the customer we serve, as well as the health and well-being of employees.

The real transition underway is not the massive automation of knowledge work. It is the redefinition of human contribution. Committing to building AI capability with employees is a win-win-win. Employees who learn AI tools, even with a foundational level of AI fluency, can improve AI output. Employees can streamline and improve processes based on what they know should be automated, not just anything that can be automated, which improves productivity, and because employees then have more time to focus on evaluating impact and advancing functionality, customer outcomes can be continually improved. Therefore, constructing opportunities for employees to expand their AI knowledge and experience builds trust and equips them to be increasingly more agile and better prepared for the future, which is the transformation needed.

Once they are AI fluent, employees can spend more time in human activities, such as consciously outsourcing, supervising, and managing the work that machines do better. This means redesigning and elevating roles to require interpretation, accountability, and empathy. It means recognizing that someone must still decide and/or supervise what questions to ask, what outcomes matter, and what trade-offs are acceptable.

AI should be used to enhance human capabilities, enabling deeper thinking, more human connection, and more responsible leadership. When implemented well, AI frees people from tactical low-value work and creates space for higher-value human contributions. This shift requires a new social contract of work, starting with leaders being empowered to organize the workload to allow for AI learning and experimentation. Once leaders are using AI, for example, to make schedules, document meetings, send and track action items, and create processes to learn from failures, they can spend more time building human connections, managing exceptions, and rewarding experimentation.

Employees must be empowered to learn AI and to participate in redesigning how work gets done. This is not a concession; it is a strategic advantage. When employees help identify what should be automated (and what should not), processes improve in ways leaders alone cannot anticipate. Productivity increases because automation is applied where it truly adds value. And outcomes improve for customers because work is redesigned with real-world context by those who know the workflows best.

The high visibility creates high accountability, which is contagious and only happens when leaders lead the transformation with integrity, transparency, and human purpose. This partnership between machines and humans will build trust within the entire system.

The Four Pillars for Humans to Work with Machines Successfully

Redesigning work in the age of AI is not a one-time initiative. It is an ongoing leadership discipline. Organizations that experience success tend to anchor their approach in four interdependent pillars: clarity, capability, governance, and purpose. These pillars enable trust and retain human connection.

Clarity

Clarity answers the most fundamental questions people are asking, often silently:

- Why are we using AI?
- What AI skills do I need?

- What does success look like?
- What will change, and what will not?
- Are experimentation and failure OK?
- How will we measure outcomes in alignment with our mission?

What good looks like in practice is leaders communicating early and often, in plain language, about the role AI will play. It is leaders explicitly connecting AI initiatives to improved outcomes for those the organization serves, rather than abstract efficiency targets. It is aligning metrics to employees' expected AI fluency and how AI augments their work to improve output. It is leaders who are honest about uncertainty and trade-offs, rather than hiding behind technical jargon. Clarity reduces fear because it replaces speculation with shared understanding.

Capability

Capability is about more than training people to use tools. It is about redefining what competence looks like when humans lead machines. It is working with employees to redefine roles and processes. The only way to learn AI is by using it.

Organizations must define AI fluency in a way that fits their context, roles, and values. Not all employees need to be AI transformative experts. However, employees should be fluent in integrating AI into workflows and shaping how processes are performed. "Delegation engineering" is a skillset just as important as "prompt engineering."

At the same time, companies must double down on leadership development that emphasizes how to lead humans who are increasingly responsible for directing, supervising, and interpreting machine output. Most jobs will be reconfigured, and new jobs will be created, so leaders must be experts in collaboration, effective communication, continuous learning, and active listening to resolve conflicts.

What good looks like in practice is investing in learning that builds both technical literacy and human skills, ensuring all employees are held responsible for curiosity that retains critical thinking. It is teaching and empowering leaders and employees to question AI-generated outputs, apply judgment, experiment, and take accountability for decisions. It is equipping leaders to coach through ambiguity rather than control through command.

Capability signals that people are trusted to learn and adapt. Even if a position is eliminated, the employee who is AI fluent will be prepared to contribute to the future of work, which benefits all employees, business, and society.

Governance

Trust is also built through guardrails, not promises.

Ethical guidelines, transparency in decision-making, and clear accountability structures are essential as AI becomes more embedded in work. People need to know how decisions are made, what technology is approved to use, what is not approved and why, how bias is addressed, and who is responsible for documenting the learning when systems fail.

What good looks like in practice is governance that is highly visible and human centered. It is leaders openly acknowledging risks and explaining how they are being mitigated. It is creating feedback loops without fear of retaliation where employees are obligated to raise concerns. Trust grows when people see that leaders are willing to slow down to get it right.

Purpose

Finally, purpose must be anchored and aligned with the entire system.

AI will inevitably change roles, identities, and workflows. AI can also prevent overwork, burnout, and workplace injuries. Leaders must actively support people through AI transitions by reinforcing why their work matters and how human contribution continues to make a difference.

What good looks like in practice is leaders consistently naming the human impact of the organization's work. It helps employees see how their judgment, human connection, and creativity improve outcomes. It is recognizing and celebrating human contribution, not just automated efficiency. Connection ensures that people feel they belong in the future being built.

A Defining Moment for Leadership

This moment demands pause, reflection, and courage.

History will not remember what AI was capable of in this era. It will remember what leaders chose to do with it. It will remember whether leaders defaulted to speed or chose responsibility, whether they prioritized short-term gains or long-term trust, and whether they treated people as costs to be reduced or contributors to be developed.

Machines with AI will not replace humanity. But AI is redefining how we work and live together. Historically, we moved from being tactical workers to knowledge workers. Now we will transform to delegation engineers using AI to augment work to improve all outcomes. The leaders who will be remembered will have taken the opportunity and met the obligation to build organizations worth belonging to, places where smart machines support human brilliance.

It's not about only machines and it is not about only humans. It is about how they work together. While human wisdom remains the most powerful intelligence, we must use our wisdom to shape AI and the mechanics of work. The only parts of a job in the future will be the human parts. Let's articulate what human contributions must continue. Human leadership has this obligation to determine how we work with machines and whether we build a future worth working and living in. Only people will know if things are improved.

References

1. Jason A. Colquitt, Brent A. Scott, and Jeffery A. Lepine, "Trust, Trustworthiness, and Trust Propensity: A MetaAnalytic Test of Their Unique Relationships with Risk Taking and Job Performance," *Journal of Applied Psychology* 92, no. 4 (2007): 909–27.
2. U.S. Bureau of Labor Statistics, *The Rise in Remote Work Since the Pandemic and Its Impact on Productivity*, Beyond the Numbers, vol. 13, no. 7 (U.S. Department of Labor, 2024), https://www.bls.gov/opub/btn/volume-13/remote-work-productivity.htm.
3. Tammy D. Allen et al., "Remote Work: PostCOVID19 State of the Knowledge and BestPractice Recommendations," *Industrial and Organizational Psychology* 17, no. 3 (2024): 337–62.
4. Bing Wang et al., "Achieving Effective Remote Working During the COVID19 Pandemic: A Work Design Perspective," *Applied Psychology* 70, no. 1 (2021): 16–59.
5. Brian Elliott, "ReturntoOffice Mandates: How to Lose Your Best Performers," *MIT Sloan Management Review*, November 11, 2024, https://sloanreview.mit.edu/article/return-to-office-mandates-how-to-lose-your-best-performers/.
6. Shaul Oreg and Yair Berson, "Leaders' Impact on Organizational Change: Bridging Theoretical and Methodological Chasms," *Academy of Management Annals* 13, no. 1 (2019): 272–307.
7. U.S. Bureau of Labor Statistics, *Effects of Remote Work on Productivity in COVID19 Era*, DPR Notice (U.S. Department of Labor, 2025).

About the Author

Stacy Proctor advises business leaders on one of the most critical challenges of our time: how to implement AI in a way that strengthens, rather than erodes, trust, performance, and human value.

A Chief People Officer with over 25 years of experience, Stacy is known for aligning people strategy with business outcomes across purpose-driven organizations, including biotech, sustainable energy, and medical technology. Her work focuses on organizational design, leadership development, and change leadership, helping companies build high-performing cultures rooted in trust.

Stacy's approach to AI is grounded in a simple belief: AI success is determined by how leaders engage their employees. She coaches mid-sized company leaders to implement AI with intentionality, empowering employees to build AI fluency, co-create what should be automated, ensuring customers feel the benefits of AI and employees trust the change.

Recognized for her practical, human-centered style, Stacy helps organizations redefine work so that machines enhance human contributions and companies have long-term success.

She holds an MBA and a BA in Sociology, along with certifications including SPHR, PHR-CA, SHRM-CP, and an "Advanced Lead with AI" Certification.

Website: https://stacyproctorconsulting.com/

LinkedIn: https://www.linkedin.com/in/stacyproctor/

CHAPTER 23

THE NEW MARKETING MACHINE: DRIVING STRATEGY AND SPEED THROUGH TEAM TRANSFORMATION

By Jenny Lopez Riley
Chief Growth Officer, AI-Forward Leader
Chicago, Illinois

Coming together is a beginning. Keeping together is progress. Working together is success.

—Henry Ford

Let me begin with one truth we cannot ignore: Marketing is being rebuilt in real time. Consumer and buyer expectations continue to rise while budgets tighten. Technology, particularly artificial intelligence (AI), is advancing faster than most organization structures can absorb. What keeps marketing leaders up at night isn't the promise of

AI itself; it's the widening gap between what our tools are capable of and how our teams are designed to deliver. Simply put, the org charts that brought us here will not take us where we need to go next.

Winning in this new era requires more than adopting new platforms. It requires rethinking how marketing teams are structured, how work flows, and how leadership translates strategy into measurable impact, fast. The leaders who succeed will not be those who optimize yesterday's models, but those who have the courage to redesign the system entirely.

The Forces Reshaping Marketing Teams

Across industries, work itself is being fundamentally redefined: where and when it happens, how it happens, and why it matters. For marketing organizations competing in an always-on, digital-first economy, understanding the forces driving this shift is essential. These forces are interconnected, reinforcing one another and reshaping expectations for talent, speed, and adaptability.

Generational shifts are one of the most visible changes. For the first time, five generations are working side by side. Millennials now represent the largest share of the workforce, with Gen Z rapidly growing behind them while Baby Boomers continue to phase out. According to US labor data, Gen Z surpassed Baby Boomers in workforce participation in 2024.[1] But this is not simply a demographic footnote, it's a leadership challenge.

Gen Z and Millennials consistently place a high priority on compensation, meaning, flexibility, and well-being. Yet fewer of them view traditional leadership roles or linear career ladders as aspirational. This signals a profound shift in how motivation, growth, and success are perceived. Incentives and structures that worked for previous generations—rigid hierarchies, legacy workflows, or tenure-based advancements—often disengage cohorts that value autonomy, purpose, and digital fluency. Marketing leaders must reconcile these worldviews if they want to harness cross-generational talent effectively.

At the same time, global marketplaces have unbundled talent from geography. Platforms such as Upwork and Fiverr, combined with cloud infrastructure from companies like Amazon, Apple, and OpenAI, allow organizations of any size to access world-class capabilities on demand. Talent procurement is no longer limited to location or full-time employment. Small, focused teams can now operate with the reach and sophistication of large enterprises by integrating specialized talent when and where it's needed.

This shift is reinforced by new ways of working. Independent work, freelancing, and side hustles have moved from the margins to the mainstream. Recent workforce analyses suggest that more than a third of US workers engage in some form of gig or project-based work,[2] and the number of people holding multiple jobs has reached its highest level in decades.[3] For marketing leaders, this means workforce strategy can no longer rely solely on permanent headcount. Agility now depends on blending full-time employees with project-based specialists who can scale up or down as market needs change.

Layered on top of this is the lasting impact of COVID-19, which permanently altered employees' relationship with work. Remote and hybrid models proved that productivity and flexibility are not mutually exclusive. As of early 2025, a majority of remote-capable roles in the US continue to operate in hybrid or fully remote arrangements.[4] More importantly, the pandemic triggered a reassessment of purpose. Many professionals now prioritize meaningful work and values alignment alongside compensation, raising the bar for leaders to articulate not just what the work is, but why it matters.

Finally, and with the greatest disruption, technology, especially AI, is reshaping capability itself. Generative AI and automation are driving the cost of knowledge toward zero. Tasks that once required teams of specialists can now be performed in seconds. But AI is not an isolated force. It compounds every other shift: flexible talent models, remote work, global competition, and rising expectations. The organizations that win will be those that integrate AI with human creativity, judgment, and strategy rather than treating it as a standalone tool.

These forces are not independent. They intersect and amplify one another. Together, they demand a new marketing machine, one designed for speed, adaptability, and continuous learning.

Designing Teams for a World That Never Slows Down

AI is no longer advancing incrementally. It is evolving exponentially. Most organizations struggle to absorb one wave of change before another is already reshaping their industry. As a result, marketing leaders must move from asking, "How do we do more?" to a more fundamental question: "How do we design teams that move at the speed the world now operates?"

This begins with recognizing the scale of reskilling required. The World Economic Forum estimates that nearly 60% of the global workforce will need reskilling or upskilling by 2030.[5] The idea that professionals can learn a fixed set of skills early in their careers and rely on them for decades is obsolete. Continuous learning is no longer a perk; it is a survival requirement.

Marketing feels this pressure acutely. Every function—content, analytics, media, customer journeys, measurement—is being redefined by AI. Roles that were once discreet are now converging. Capability decay has become as dangerous as competitive pressure. Teams that are not designed to renew skills continuously will fall behind faster than traditional metrics can detect.

The organizations that succeed will treat capability transformation as a competitive advantage, not a reactive cost. They will design teams that anticipate change, rather than scrambling to respond to it.

Operating on Outdated Hardware

Here's the metaphor to remember: Most marketing teams today are operating with the equivalent of a flip phone. We expect real-time personalization, always-on engagement, rapid experimentation, and measurable impact. Yet the underlying structures—org charts, workflows, decision rights, and skill models—have not kept pace. Decision cycles stretch from weeks to months in a market that now moves in days.

This is not a talent problem.

It is not an effort problem.

It is a design problem.

We keep layering new software on top of outdated systems, hoping tools will compensate for structures that were never built for this level of speed and complexity. They can't. The world requires a smartphone, not because it is faster in isolation but because it is designed for speed, integration, and constant change. Until we upgrade the hardware—how workflows are created, how decisions are made, how skills are combined—marketing will continue to feel overwhelmed. Smartphones work because the hardware, operating systems, and applications are built together intentionally.

The result is predictable. Teams feel overwhelmed. Leaders feel frustrated. Execution feels slower every year, even as tools get better.

The new marketing machine does not start with more tools. It starts with rebuilding the systems and teams those tools run on.

Where AI Drives the Biggest Impact

One of the first things I recommend every marketing leader should do is assess every role, not to reduce headcount, but to identify where AI can eliminate wait time. The real value of AI in marketing is not doing more work, it's eliminating friction.

AI delivers its biggest impact in three areas.

First, repetitive production. AI can generate first drafts, variations, SEO optimization, template outputs, and research at scale, freeing teams to focus on quality, storytelling, and brand judgement. This isn't about replacing creativity; it's about protecting it.

Second, coordination and operations. Marketing speed is often constrained by handoffs, approvals, and administrative work. AI can manage calendars, workflows, and documentation. By offloading this work, teams regain time for strategy, decision-making, and orchestration, the activities that actually move the business forward.

Third, analysis and optimization. AI can synthesize massive datasets in real time, generating insights continuously rather than quarterly. This changes the cadence of marketing from reactive to adaptive.

AI doesn't make teams better by doing the work for them. It makes them better by removing delays between them. And when speed becomes the advantage, that difference is everything.

The Rise of AI Agents and the Shift in Roles

I recently asked a group of 70 CMOs what they are doing with AI agents. No hands went up. I must admit, I'm still learning this space myself. Here's what I do know. The paradigm is shifting from manual human effort to humans managing intelligent agents that operate at scale.

AI agents are not just assistants. They think, plan, and act. They break goals into tasks, operate across tools and platforms, and improve over time. This is the most powerful wave of automation we've ever seen.

Automation isn't new. What's new is agency. AI agents can:

- Plan and execute multi-step workflows
- Use tools, APIs, and data sources
- Remember context and learn from feedback

That changes how work gets done.

As an example, we can now create in-house agents that are deal scouts for M&A teams. You task the agent to identify 12 US-based AEC firms with $10 to $15 million in revenue and 15%+ EBITDA.

The agent researches the market, identifies and vets targets, flags risks, delivers structure summaries, and builds from there. The end result is a 91/100 quality score with 20 to 30 minutes of human review. The same thing can be done for an email or social media agent. What used to take weeks now takes minutes.

For marketing, this doesn't eliminate roles, it elevates them. Titles built around manual execution will shrink or evolve. The future belongs to roles that design systems, direct intelligence, and shape outcomes.

Roles That Shrink or Fundamentally Evolve

Current Role	New Role	Evolution
Content Specialist	Director of AI Content	Junior copywriters become AI editors or narrative architects, guiding voice, quality, and story instead of drafting from scratch.
Production Designer	Creative Systems Managers	The role moves from making assets to designing the systems that generate and adapt them.
SEO Specialist	Search Intelligence Analysts	There is less keyword tuning and more interpretation of intent, signals, and search behavior across platforms.

Marketing Coordinator	Marketing Systems Strategist	One of the most important emerging roles. This role designs and manages the end-to-end flow of work, aligning AI agents, people, systems, and content, so everything moves in the right sequence, at the right time, with minimal friction.
Social Media Manager	Social Intelligence Strategist	AI can draft, schedule, tag, repurpose, and report. Humans interpret culture, nuance, influence, and emerging signals, things AI cannot fully understand.

This is what the new marketing machine demands.

Orchestration: The Real Source of Speed

Henry Ford once said, "Coming together is a beginning. Keeping together is progress. Working together is success."

Think of Formula 1. Races are often won in the pit lane, where perfectly rehearsed teams execute with precision under pressure. Marketing works the same way. When humans and AI are orchestrated effectively—each doing what they do best—everything moves faster.

How is that possible? Not individual brilliance. Orchestration. Every role is precise. Every moment is rehearsed. The workflow is simple, clear, and optimized for speed.

That is the lesson for modern marketing teams.

Enablement plus orchestration is the competitive advantage. Design, not effort, determines speed.

The Responsibility of Leadership in the AI Era

We often describe AI as the great equalizer. In theory, equal tools should create equal opportunity. In reality, they don't.

AI is being built on existing structures, and without intention, it amplifies inequality. Entry-level roles, the very jobs that once allowed people to learn, earn, and advance, are often the first to be automated. Access to technology, training, and mentorship is uneven, widening gaps rather than closing them.

This is where leadership matters most.

The only truly sustainable advantage, regardless of company size, is talent. Talent turns tools into value. AI is a force multiplier, not a differentiator. Most companies will have access to the same technologies. What separates winners is a workforce that is well trained, motivated, and supported.

For every dollar invested in AI, multiples must be invested in people training, upskilling, and development.

Leadership is not defined by how high we climb, but by how many people we bring with us.

In this era of uncertainty, our responsibility extends beyond performance metrics. It is about ensuring opportunities, guidance, and a real path forward. We rarely know someone's full story: the barriers they've faced, the rooms they were never invited to, the weight they carry.

How we lead matters more than ever.

Lead with kindness.

Lead with grace.

Lead with a growth mindset.

AI will change our world. But we decide what kind of world we build with it. And the future will not remember us for the tools we adopted, but for the people we lifted as we climbed.

References

1. U.S. Department of Labor, Employment and Training Administration, "Trendlines: August 2024 Edition," *Trendlines Newsletter*, August 2024, https://www.dol.gov/sites/dolgov/files/ETA/opder/DASP/Trendlines/posts/2024_08/Trendlines_August_2024.html.
2. Ibrahim Zaghw, "Gig Economy Statistics for 2025," *Finmasters*, updated July 5, 2025, https://finmasters.com/gig-economy-statistics/.
3. Terry Lane, "More Americans Are Working Multiple Jobs Than at Any Time Since 1999," *Investopedia*, January 2026, https://www.investopedia.com/more-americans-are-working-multiple-jobs-than-at-any-time-since-1999-11880331.
4. Great Place to Work, "Remote Work Productivity Study Finds Surprising Reality: Two-Year Study," *Great Place to Work Blog*, accessed March 2026, https://www.greatplacetowork.com/resources/blog/remote-work-productivity-study-finds-surprising-reality-2-year-study.
5. World Economic Forum, "Future of Jobs Report 2025: 78 Million New Job Opportunities by 2030 but Urgent Upskilling Needed to Prepare Workforces," press release, January 8, 2025, https://www.weforum.org/press/2025/01/future-of-jobs-report-2025-78-million-new-job-opportunities-by-2030-but-urgent-upskilling-needed-to-prepare-workforces/.

About the Author

Jenny Lopez Riley's passion for mentorship is what drives her to create high-performing teams and leaders. With more than 20 years of experience accelerating revenue through digital transformation and technology adoption, she has led brand transformation, integrated marketing, and enterprise growth initiatives for complex, global organizations. She is known for building high-performing teams by investing in people as intentionally as she invests in strategy.

As chief growth officer at RTM Engineering Consulting, Jenny leads enterprise growth, innovation, marketing, communications, and client experience, aligning data, technology, and talent to drive competitive advantage. Throughout her career, she has driven automation and advanced analytics, extending that work into AI-powered workflows to streamline operations, accelerate decision-making, and elevate client engagement.

A recognized thought leader and speaker, Jenny is deeply committed to lifting others, fostering inclusion, and developing the next generation of leaders in an AI-powered world.

Email: jloriley53@gmail.com

LinkedIn: www.linkedin.com/in/jenny-riley

CHAPTER 24

A SELF-ADAPTIVE APPROACH TO BUILDING CREATIVE INTELLIGENT ROBOTS

By Paul Robertson, DPhil
AI Researcher
Boston, Massachusetts

I visualize a time when we will be to robots what dogs are to humans, and I'm rooting for the machines.

—Claude Shannon

Since the earliest days of artificial intelligence, researchers have debated what intelligence actually is. Alan Turing's seminal 1950 paper, "Computing Machinery and Intelligence," reframed the question operationally through what became known as the Turing Test. Since then, different schools of thought—symbolic AI, connectionism, probabilistic modeling, and more recently large-scale deep learning—have each claimed to offer the definitive path forward. Yet no universally accepted definition of intelligence exists, nor is there agreement on how to measure it (see Legg & Hutter, "Universal Intelligence: A Definition of Machine Intelligence," 2007). Even within biology and comparative cognition, intelligence across species remains contested (see Shettleworth, *Cognition, Evolution, and Behavior*, 2010). The human brain is frequently treated as both proof of intelligence and the standard against which it is judged.

Learning

Turing, in his aforementioned paper, described what he saw as an artificially intelligent system as one that started out rather like an empty notepad to which learned experiences are added. Clearly, learning was an important part of intelligence for Turing, and despite decades of research that ignored learning, few would doubt its importance today. Today's deep learning systems, reinforcement learning systems, and large language models do not learn in an interesting sense. They all depend upon offline learning from massive corpora, or in the case of reinforcement learning, a massive number of episodes. We want our robots to learn from interacting with their environment throughout their service life.

Like perception, learning can be viewed as an active process. We learn less by observing the world pass by than by actively interacting with our environment. Learning is most effective when we are engaged in the process. This is sometimes termed "active learning," the idea that an agent learns most efficiently by selecting the experiences from which it learns, rather than passively receiving pre-collected data (Settles, "Active Learning Literature Survey," 2009).

Intelligence Is Not a Uniquely Human Trait

Early AI research targeted tasks that appeared to require uniquely human intelligence, such as playing chess. Capabilities such as computer vision (CV) and natural language (NL) processing were also identified as central, since they enable interaction with the world. Particular emphasis was placed on higher-level reasoning, abilities often assumed to distinguish humans from other animals. The desire to see humans as unique is little supported by the evidence. As those old discriminations have fallen away, we have come to observe that there is little that distinguishes us from animals, except perhaps that we are more capable in certain domains. We are certainly more prolific builders and users of tools than other animals, but other animals are known to build and use tools (see Whiten et al., "Cultures in Chimpanzees," *Nature*, 1999). Perhaps the only domain where humans can really claim to be unique is in the use of structured language (Hauser, Chomsky & Fitch, "The Faculty of Language: What Is It, Who Has It, and How Did It Evolve?," *Science*, 2002). It is perhaps not surprising, therefore, that so much emphasis has been placed on language and other uniquely human competencies like the games of Chess and Go that predominated early AI research.

Most of what we are we share with animals. We have evolved to depend less on innate capabilities and more on learning (see Deacon, *The Symbolic Species*, 1997).

Robotics and Grounding

In robotics, it could be argued that we have focused almost exclusively on innate capabilities and very little on learning. We have, especially recently, limited learning to be from large corpora, and for LLMs from the written word. We learn what is meant by a beach ball by manipulating one. We know what it feels like in our hands, how it moves when we throw it, how it is compressible. We attach a label to it, "beach ball," and we can describe that it is often played with on the beach, but that knowledge is not grounded in manipulation, which puts robotics at a serious disadvantage. The problem of connecting symbols to their real-world referents through sensorimotor experience is known as the symbol grounding problem (Harnad, "The Symbol Grounding Problem," *Physica D*, 1990). As impressive as LLMs are in producing well-written prose on any subject they have been trained on, there is no sense in which any of that is grounded in experience. We would argue for the importance of grounded understanding in any circumstance, but clearly in robotics, manipulation is paramount, and ungrounded knowledge is useless for manipulation. Hand-generated procedures for manipulation will disappear just as hand-described recognizers for objects already have in computer vision.

We like to think that we are unique in having reasoning capabilities; however, evolutionary history tells a different story. Many animals possess highly developed sensory systems that are essential for survival. Research in comparative cognition has repeatedly demonstrated sophisticated planning and problem-solving in non-human species (see Clayton & Dickinson, "Episodic-like Memory During Cache Recovery by Scrub Jays," *Nature*, 1998; Emery & Clayton, "The Mentality of Crows," *Science*, 2004). Natural language, by contrast, is a relatively recent development in evolutionary terms. Over time, it has become clear that describing objects verbally is not an effective way to build robust recognition systems. Human introspection is notoriously unreliable, a point long emphasized in cognitive psychology (Nisbett & Wilson, "Telling More Than We Can Know," *Psychological Review*, 1977). When asked to explain how we recognize something, we often provide incomplete or inaccurate accounts. Object recognition systems built from encoded human descriptions similarly tend to be brittle.

Neuroscience and Vision

Neuroscience has devoted extensive effort to understanding how the brain processes visual information. Foundational work by Hubel and Wiesel (1962) demonstrated hierarchical feature processing in the visual cortex, and later predictive-processing accounts emphasized the role of feedback and expectation in perception (see Friston, "The Free-Energy Principle: A Unified Brain Theory?", *Nature Reviews Neuroscience*, 2010). Insights from this work have influenced the development of modern computer vision systems. Human intelligence serves both as our default definition of intelligence, so ignoring what neuroscience has revealed about the brain would be unwise. Whatever intelligence we create will be compared to human intelligence.

For many years, vision was often described as a "bottom-up" process. This simplification persists in some discussions. In reality, perception reflects an interaction between bottom-up sensory input and top-down expectations (Gilbert & Li, "Top-down Influences on Visual Processing," *Nature Reviews Neuroscience*, 2013). Our eyes move rapidly through saccades toward objects of interest, guided by both stimulus and intention (Yarbus, *Eye Movements and Vision*, 1967). Perception is, therefore, not passive reception but active interpretation.

Importantly, there is no compelling evidence that this top-down modulation is driven primarily by language. Animals without structured language are capable of extracting essential information from scenes and acting upon it. Language may provide post hoc explanations, but it does not appear to be the mechanism that generates perceptual interpretation.

The strong emphasis on natural language in AI may stem in part from Turing's proposal that conversational indistinguishability could serve as a test of machine intelligence (Turing, 1950). This framing reinforced the belief that language defines intelligence. Yet intelligence likely emerged through evolutionary pressures favoring survival in complex environments. It is distributed, in varying degrees, across species. Language became increasingly important for humans, but it arrived late in evolutionary history. To assume that a language model represents the culmination of 70 years of AI research risks overlooking much of what is known about brain organization and its evolution (see Dehaene, *How We Learn*, 2020).

Intelligence becomes visible in animals when they can perceive their environment and act within it. Even the word "emotion" derives from the Latin *emovere*, meaning "to move out" or "to stir up," reflecting the close relationship between internal states and action (Damasio, *Descartes' Error*, 1994). Robots, too, are machines designed to act in the world. A focus on action, rather than language alone, may, therefore, provide a more productive foundation for artificial intelligence.

Although we can learn much from the brain, our goal need not be imitation. Nature made particular implementation choices shaped by biological constraints. Artificial systems need not replicate those choices. Inspiration, rather than imitation, may offer the most practical path forward (Brooks, "Intelligence Without Representation," *Artificial Intelligence*, 1991).

Creativity for Autonomous Robots

Autonomous robots can operate in many settings. In structured environments—factories or warehouses—the surroundings can be engineered to accommodate robotic limitations. However, robots designed for open or unpredictable environments face a different challenge. They must learn as they go. They must resolve problems not anticipated by their designers. Such robots require creativity: the ability to generate, evaluate, and execute novel solutions. They should be able to imagine the consequences of a proposed plan before executing it. Cognitive scientists increasingly describe this capacity as mental simulation or internal modeling. Josh Tenenbaum and colleagues have proposed that human cognition relies on structured generative models that function much like a "game engine in the head," allowing intuitive physical reasoning and counterfactual simulation (Tenenbaum et al., "How to Grow a Mind," *Science*, 2011; Battaglia, Hamrick & Tenenbaum, "Simulation as an Engine of Physical Scene Understanding," *PNAS*, 2013). This ability to simulate potential outcomes before acting reduces risk and increases efficiency. When we imagine a tower of blocks collapsing before we build it or foresee danger before taking a step, we are relying on such internal predictive models.

Consider a squirrel attempting to reach a bird feeder designed to be inaccessible except to birds. Observing its ingenuity reveals intelligence expressed through planning, experimentation, and persistence. Evolution has rewarded such capacities because they increase survival. Symbolic artificial intelligence provides tools for addressing this type of problem. Classical work on planning systems (e.g., STRIPS; Fikes & Nilsson, "STRIPS: A New Approach to the

Application of Theorem Proving to Problem Solving," *Artificial Intelligence*, 1971) demonstrated how symbolic representations enable structured reasoning about actions and goals. At the intersection of neural and symbolic approaches lies the possibility of flexible systems in which high-level symbolic plans are executed by neural components that handle perceptual and motor complexity. When a plan fails, the robot can analyze why it failed, adjust components, or search for alternatives. In doing so, it adapts prior knowledge to new circumstances. This process defines self-adaptation.

Creativity, at its core, is structured problem solving. Symbolic reasoning enables abstraction by suppressing irrelevant detail. Children illustrate this principle. Through play, they acquire competencies and abstract reusable strategies, such as stacking objects to reach higher locations (Vygotsky, *Mind in Society*, 1978). Once internalized symbolically, such strategies can be applied in novel contexts. When a squirrel stacks objects to reach food, it applies an abstract strategy—achieving height through construction—to a specific situation. It chains together learned competencies into a novel configuration. Tenenbaum's "intuitive physics" framework suggests that such reasoning relies on internal generative models capable of simulating object dynamics (Battaglia et al., 2013). A robot equipped with comparable abstraction and simulation capabilities could display similar creativity.

The process can be summarized:

1. Learning through exploration
2. Symbolic abstraction
3. Meta-level search
4. Simulation and refinement
5. Adaptation and consolidation

Symbolic reasoning enables consideration of classes of solutions. Simulation and experimentation refine instantiations. This approach has been applied to scene understanding in Robertson's grounded reflective agent vision architecture (GRAVA), enabling multi-level reasoning to generate and adapt interpretation programs. By combining deep learning for object recognition (LeCun, Bengio & Hinton, "Deep Learning," *Nature*, 2015), reinforcement learning for skill acquisition (Sutton & Barto, *Reinforcement Learning*, 2018), and reflective symbolic reasoning for adaptive planning, robots may achieve robustness in novel environments.

Abstraction and meta-level reasoning provide a way to efficiently reason at a high level covering infinite potential solutions. Breaking a rigid object by hitting it with a solid object provides an abstract solution that can be instantiated as opening a nut by hitting it with a rock or a hammer.

The GRAVA architecture uses computational reflection to build a representation of a complex scene. Computational reflection—the capacity of a system to reason about and modify its own inferential processes—was introduced by Brian Cantwell Smith in his foundational work on procedural reflection in programming languages (Smith, "Procedural Reflection in Programming Languages," MIT PhD dissertation, 1982; Smith, "Reflection and Semantics in Lisp," *POPL*, 1984). A reflective tower allows very abstract strategies for understanding a scene to be instantiated as very specific solutions by refinement through layers. Individual instances that may be encountered through exploration can be arbitrarily abstracted to support future creative thinking. Future uses of these abstractions can lead to bifurcation of abstractions—compare this with Piaget's assimilation and accommodation (Piaget, *The Origins of Intelligence in Children*, International Universities Press, 1952).

Conclusions

Self-adaptation provides a coherent framework for integrating artificial neural networks with symbolic reasoning in robotic intelligence. Until recently, it was widely assumed that complex problem-solving required language. However, research by Evelina Fedorenko and colleagues has shown that individuals with severe aphasia can perform complex non-verbal reasoning tasks successfully

(Fedorenko & Varley, "Language and thought are not the same thing: Evidence from neuroimaging and neurological patients," Annals of the New York Academy of Sciences, 2016). Functional MRI studies demonstrate that language-selective brain regions are not activated during logical or mathematical reasoning in neurotypical participants (Fedorenko, Duncan & Kanwisher, *PNAS*, 2013). Instead, such tasks recruit a domain-general "multiple-demand" network (Duncan, "The Multiple-Demand (MD) System of the Primate Brain," *Trends in Cognitive Sciences*, 2010). These findings suggest that structured reasoning operates independently of language-processing circuits. Language may primarily serve communicative and social functions. The phenomenon of inner speech remains under investigation (Vygotsky, *Thought and Language*, 1934; Fernyhough, *The Voices Within*, 2016).

The GRAVA architecture illustrates how meta-reasoning can be implemented in artificial systems. It supports hierarchical reasoning levels, though in practice three appear sufficient: contextual goal maintenance, adaptive refinement, and concrete execution planning. Originally, GRAVA relied on special-purpose recognizers. Modern deep learning recognizers would likely improve robustness, but they face the problem of catastrophic forgetting (McCloskey & Cohen, "Catastrophic Interference in Connectionist Networks," *Psychology of Learning and Motivation*, 1989; Goodfellow et al., "An Empirical Investigation of Catastrophic Forgetting in Gradient-Based Neural Networks," 2013). Addressing continual learning remains essential for self-adaptive systems operating in dynamic environments (see Parisi et al., "Continual Lifelong Learning with Neural Networks: A Review," *Neural Networks*, 2019).

Taken together, these considerations reinforce the central claim of this chapter: intelligence—biological or artificial—is grounded in perception, action, abstraction, and adaptation. Language may extend these capacities, but it does not constitute their foundation. A self-adaptive neuro-symbolic architecture offers a scientifically grounded pathway toward robust and creative robotic intelligence.

About the Author

Dr. Paul Robertson, born in England, discovered his passion for problem-solving through chess, even developing several chess-playing programs during his pursuit of a computing science degree. But it was artificial intelligence and computer vision that ultimately captured his imagination.

He pursued that curiosity to the University of Oxford, where he completed a D.Phil in Self-Adaptive Architecture for Image Understanding and helped organize the inaugural International Workshop on Self-Adaptive Systems (IWSAS), a series that eventually grew into the Self-Adaptive and Self-Organizing (SASO) conference.

His research career took him to MIT and BBN Technologies before he founded his own research company dedicated to the challenges of AI and robotics. Inspired by nature, Dr. Robertson is passionate about developing energy-efficient AI for robots and advocates for uniting symbolic AI with artificial neural implementations through self-adaptive approaches, a challenge he sees as one of the field's most important frontiers.

Email: paulr@dollabs.com

Website: www.dollabs.com

LinkedIn: https://www.linkedin.com/in/probertson/

CHAPTER 25

FROM THREAT TO THRIVE: INTRODUCING AI THRIVE HUMAN-AI PERFORMANCE FRAMEWORK

By Dennis Sanchez, PsyD
Founder of Perform/AI, Chief Human-AI Architect
Canton, Georgia

My mission in life is not merely to survive, but to thrive.
—Maya Angelou

It is somewhat paradoxical that artificial intelligence (AI), the most advanced technology in our era, is bringing out the "beast" in us. Just as our early, primitive ancestors automatically responded to threats with a "fight or flight" instinct (MacLean, 1983), AI often triggers an instinctive negative reaction, rooted in fear, that an individual may experience consciously or unconsciously when they interpret this technology as a threat to their well-being (Panksepp, 2004).

Resistance to adopting new technology is hardly new for organizations. For example, when cloud computing began gaining mainstream traction in the early 2000s, many CIOs chose not to migrate because they considered it a threat to their organizations' security. Similarly, recent research indicates that many workers view AI as a workplace threat and actively resist or refuse to use AI tools (Lindzon, 2025).

Negative reactions to AI are also affecting one of the key metrics used to measure an organization's return on investment (ROI) from AI investments: AI adoption. AI adoption refers to the cognitive and behavioral shift that occurs when users not only use the technology frequently but also form a new relationship characterized by trust, collaboration, and an emotional bond (Sanchez, 2024).

Achieving this level of AI adoption is fundamentally a new way of thinking about the relationship between people and technology, areas I've been deeply engaged in for more than a decade. To move forward, the first step is to stop treating AI as if it were a past technology, such as personal computers or traditional software, which were never perceived as existentially threatening.

More importantly, we need to redefine what it requires of each of us to tune out fear-centric narratives surrounding AI and instead focus on what becomes possible when we intentionally integrate AI into our lives, a key requirement for AI adoption. Finally, we each need to discover the positive side of AI and operate permanently in the Thrive Zone—the third stage of the AI Thrive HumanAI Performance Framework that we will explore together.

Redefining Thrive in the AI World

Before we get too deep into a discussion of the definition of "thrive" in an AI context, it is important to recognize that this term is also widely used in psychology, education, business, and the workplace. According to the Merriam-Webster Dictionary (n.d.), thrive is a verb that means to grow vigorously: flourish, to gain in wealth or possessions, and to prosper toward or realize a goal despite or because of circumstances.

There are at least two definitions of "thrive" used in psychological contexts. The first version uses "thrive" to describe a state that follows a person overcoming an adverse condition and rebounding even stronger, thereby emphasizing resilience (O'Leary & Ickovics, 1995). The second definition is most often associated with human flourishing, a psychological state characterized by high well-being and optimal functioning (Seligman & Csikszentmihalyi, 2000).

In education, the term "thrive" is most often associated with student success and well-being. Thriving students are described as fully engaged socially, emotionally, and academically, and as moving toward their goals (Schreiner, 2010). Thrive in the business can mean anything from creating sustainable performance and organizational resilience, to workplace wellness programs and cultures (Spreitzer & Porath, 2012).

Lastly, the definition for thrive in the workplace refers to a positive psychological state characterized by a shared sense of vitality and learning (Kleine & Zacher, 2019). It is important to note that positive associations have been found between thriving and psychological capital (hope, efficacy, resilience, and optimism), proactive personality, positive affect, and relational characteristics (Kleine & Zacher, 2019).

Among this collection of definitions, I propose that "thrive" in the AI context is both comparable and distinct from the nuanced terms used in psychology, education, business, and the workplace. "AI Thrive" is similar in that we are building on the traditional characteristics of Thrive in how it impacts an individual cognitively, emotionally, and physiologically, but different in that we are redefining the Thrive taxonomy to include the relationship between humans and AI. Let us now discover how the AI Thrive HumanAI Performance Framework can help all AI users reach this desired state.

Overview of the AI Thrive Human-AI Performance Framework

The purpose of the conceptual AI Thrive Human-AI Performance Framework (AI Thrive) is to show how any AI user can navigate a journey that begins with their first AI interaction and progresses over time to reach the desired state for working with AI, defined as the "Thrive Zone." Figure 1 illustrates the three phases of AI Thrive, including the cognitive, emotional, physiological, and behavioral dimensions of each phase.

Figure 1

The AI Thrive Human-AI Performance Framework

STEP 1	SURVIVE ZONE Primate Reaction
Cognitive:	Instinctual reactions, little cognitive interference
Emotional:	Fear and anger
Physiological:	Negative feelings in the body
Behavior:	Infrequent use or rejection, anxious and avoidant
STEP 2	**STRIVE ZONE Skill Building**
Cognitive:	Rational part of the brain involved
Emotional:	Negative and positive; co-regulation with others
Physiological:	Negative and positive feelings in the body
Behavior:	Commitment to learn, cautious curiosity
STEP 3	**AI THRIVE ZONE Human-AI Collaboration**
Cognitive:	Higher-level cognitive control & judgment
Emotional:	Trust, psychological safety, seeking, caring
Physiological:	Mixed emotional responses: negative, positive & neutral
Behavior:	Proactive collaboration with AI, satisfaction from interaction

Each of these dimensions represents a shift in how people work and feel in an AI environment. They are interrelated and change as the individual moves forward or backward because the transition to AI Thrive is not always linear. The framework is best understood through this real-life scenario, which follows a fictional character, Paula, transitioning through the three zones.

Step 1: The Survive Zone

The Survive Zone is where it all starts in our relationship with AI. Paula is an accomplished salesperson, but she has only heard of ChatGPT-3.5, and after trying it at home, she walked away feeling, "It's no big deal." The next day at work, she learns from her boss that her organization is adopting Microsoft 365 Copilot and that her department is going to be the first one "to be automated." Like an animal sensing danger from something uncertain, the news triggers Paula's built-in survival system, and she immediately calls her spouse to tell him it's only a matter of time before she is replaced. She has a sick feeling in her stomach as she scans her computer for the most recent copy of her resume.

Analysis: Based on the diagram, Paula had the equivalent of a primal reaction to the AI news. There was nothing rational about her response because her fears triggered negative emotions in the primitive part of the brain, where there is no debate, just action to flee and update her resume.

Step 2: The Strive Zone

After talking with peers who were sharing their interpretations of AI, the social connection calmed Paula, and her thoughts about AI shifted to the newer (in evolutionary terms) part of the brain, where it operates more rationally. At times, she felt negative about AI, then caught herself feeling positive, sometimes simultaneously. The positive change in how she felt physically and emotionally about AI was encouraging, and she decided to use Copilot more frequently, even though sometimes it required a lot of patience. Additionally, it was impressive that Copilot suggested several new ways to market to her customers without her request.

Analysis: The Strive Zone is the transition from early interactions with AI, and the ultimate AI Thrive state. This is where utilization begins to increase and why Paula started shifting her mindset from closed to open and curious. There was also a small reward when Copilot proactively, without solicitation, offered a sensible suggestion. Connecting with others when we face uncertainty is also how we are "wired" to deal with fear, and these conversations helped Paula feel more positive overall.

Step 3: The AI Strive Zone

In many ways, Paula worked hard to earn a place in the AI Strive Zone, which became possible when AI was no longer a threat to her well-being. In fact, she started looking forward to her interactions with Copilot, including taking it home to explore its capabilities without interruptions. Even when Copilot made mistakes, Paula was patient with the process, to the point that she felt obligated to learn how to get the best results. Trust developed, and as trust grew, she increased her utilization, leading to more synchronous dialogues in which she paid attention to all Copilot's suggestions. There were still moments when Paula felt negative about Copilot, but most of the time she was extremely positive or, at worst, neutral. Her longer sessions with Copilot were starting to pay off in higher sales, prompting Paula to continue seeking new ways Copilot could improve her overall well-being, including making more money and getting a promotion.

Analysis: Paula's transition into the AI Thrive can be characterized by a handful of key moments that changed the course of her relationship with AI, which was manifested in Microsoft 365 Copilot. First, she overcame many of her fears about AI's entry into her organization by sharing them with peers who were going through the same process. The psychological term for getting together to stabilize one another is co-regulation, but a more familiar term is psychological safety, which we are wired to seek when we are upset. Our ancient ancestors practiced co-regulation frequently, so maybe we should follow their example.

The second impactful moment in Paula's AI Thrive journey was when she changed her mindset from "I don't even want to try it" to "maybe I should give it a try." According to Carol Dweck (2016), this is called moving from a "fixed mindset" to a "growth mindset," and Paula's dramatic shift in thinking opened possibilities she likely would have never experienced, like keeping her job.

The third moment, when Paula began looking at Copilot as a collaborator, literally opened the door to the AI Thrive Zone for her to enter, and the subtle shift in how she worked with Copilot became a game-changer. Copilot was no longer "just a tool," it was an opportunity for Paula to reach her full potential in ways she had never imagined.

Conclusion

Paula's AI Thrive journey is a classic example of how easy it is for an AI user to move from struggling with AI in the Survive Zone, to being rewarded for spending the time to get to know AI in the Strive Zone, to experiencing well-being and a sense of the potential AI has to make our lives better in the AI Thrive Zone. I also want to bring to your attention a key point that has been embedded in the AI Thrive Human-AI Performance Framework: We should focus more on how AI affects our lives cognitively, emotionally, and physiologically, and discuss these topics openly and transparently.

My experience working with the public is that many people remain stuck in the Survive Zone and do not know how to get out. I would like you to please consider this question: "Am I in the Survive Zone, Strive Zone, or the Thrive Zone?" Celebrate if you are operating in the Thrive Zone. Socialize what is working for you and help others who are still struggling in the Survive Zone to get comfortable.

If you are in the Strive Zone, keep up the great work. You are so close to finding the positive side of AI, and all you are probably missing is more one-on-one time with your favorite AI solution. If you look around and see yourself operating in the Survive Zone, do not despair. You are in great company with many others, but now you have something they do not. You understand that by changing your mindset, motivating yourself to practice a little each day, and being patient as you build a long-term relationship with AI, you are on your way to forming a collaborative relationship with AI, where a new level of confidence, self-fulfillment, and success is waiting for you. Best of luck on your journey!

About the Author

As the founder of Perform/AI, Dr. Dennis Sanchez is dedicated to empowering individuals, workers, entrepreneurs, and small business owners to thrive in an AI-powered world. To further this mission, he developed ARC/AI, an eight-week, behavior-based self-upskilling program designed to help solo entrepreneurs and professionals rapidly adopt AI tools such as Microsoft 365 Copilot. ARC/AI enables participants to harness their self-motivation and AI technology to launch and manage successful one-person businesses.

With a career focused on global technology and change management, Dr. Sanchez has studied and applied insights into how people interact with technology in the workplace, culminating in his thesis on the determinants of Microsoft Cloud adoption. His passion for continuing to explore the human-technology relationship led to doctoral research supporting the theory that positive emotions significantly enhance the effective use of Generative AI. This research now fuels an ongoing commitment to advancing human-AI performance and understanding its impact.

Email: drdennissanchez@performai.ai

Website: www.performai.ai

LinkedIn: linkedin.com/in/dennis-sanchez-atlanta

CHAPTER 26

THE LEADERS WHO WILL GET AI RIGHT (AND THOSE WHO WON'T)

By Michael Schmid, PhD
Founder of Fortify Intelligence; Scaling AI Safety
Boston, Massachusetts

The test of a design is not in the laboratory but in the marketplace.
—Peter Drucker

Part I—Why Many Companies Haven't Thrived

The Runway Reality

Over the last decade, Artificial intelligence (AI) has accumulated a peculiar kind of momentum. Investment has accelerated, valuations have expanded, and each successive generation of models has appeared more capable than the last. To anyone watching the field from the outside, the trajectory looks inevitable: smarter systems, broader adoption, and a steady transformation of entire industries.

Yet a closer look reveals a mismatch that is becoming increasingly difficult to ignore.

The market value of AI companies continues to climb, often based on expectations of near-term deployment at scale. But revenue curves rarely follow the same trajectory (Figure 1). Systems that perform impressively in demonstrations stall when moved into production. Pilots extend indefinitely. Promised capabilities remain "one year away" for years at a time. Quietly, many projects are abandoned altogether.

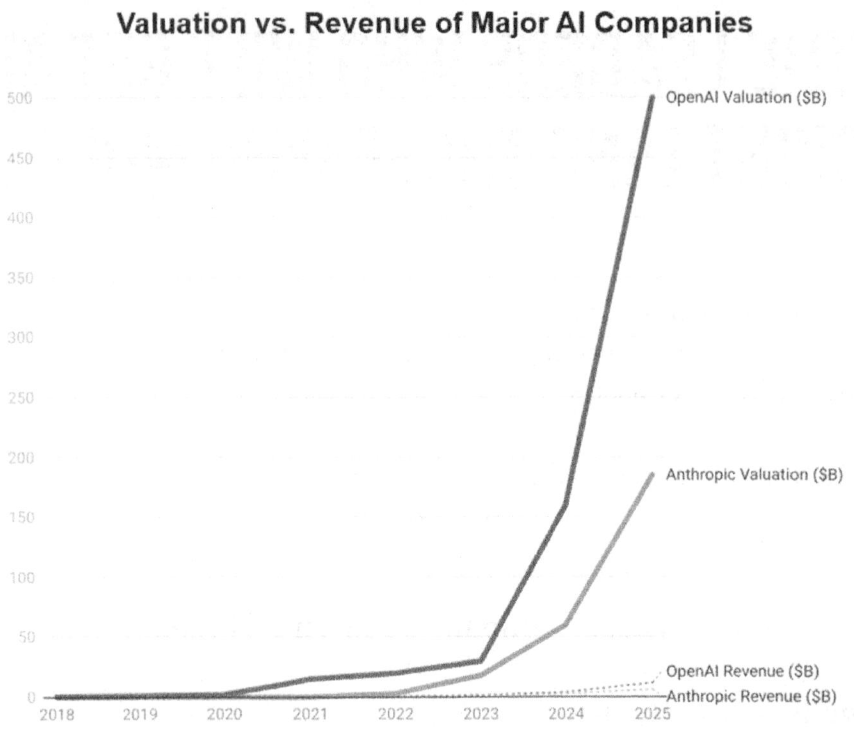

Figure 1. Valuation vs. revenue of frontier AI labs (Source: Author's compilation based on publicly reported data on OpenAI and Anthropic valuations and revenues, including company disclosures and media reporting (e.g., Yahoo Finance, Reuters, Bloomberg, The Information, 2021–2026). Credit: Datawrapper)

The pattern is widespread enough to suggest that something more fundamental is happening than a handful of execution mistakes. If capability were the only constraint, the steady improvement of models should translate directly into steady deployment success. Instead, the industry has produced a growing collection of technically impressive systems that prove far harder to operationalize than anticipated.

The result is a widening gap between promise and reality that makes the axes have two new meanings: One axis represents what these systems are believed to be capable of: widespread deployment across tasks and domains. The other axis represents operational performance: what AI can actually be trusted to do in the real world. Between them lies what might be called the AI deployment gap, the space where most AI initiatives struggle or stall (MIT-affiliated research reports that ~95% of enterprise generative-AI pilots fail to produce measurable ROI or P&L impact).

This gap is manageable in the short term, but it creates the runway problem: Companies advance on funding and must scale before that capital is exhausted. They rely on rapid deployment, but when the deployment gap persists, they approach the end of the runway.

Understanding the deployment gap requires looking beyond model capability itself and examining how organizations attempt to scale AI in practice.

The AI Scaling Trap

From a business perspective, the process tends to follow a predictable sequence.

A company develops a prototype model. The prototype performs well in controlled tests and internal demonstrations. Early customers are impressed. Funding follows. The organization expands with the expectation that scaling deployment is primarily a matter of the company's capacity.

Following promising pilots, the system is introduced into increasingly realistic environments. At first, progress appears smooth. But as exposure grows, unexpected failures begin to surface.

Edge cases appear that were not anticipated during development. Performance varies across contexts that previously seemed similar. Seemingly minor conditions produce disproportionate effects. The team responds as most engineering teams would: They collect examples of failures, retrain the model, and try again.

For a while, this works. Then, with scale, more and more failures emerge. More examples are collected. More patches are added. The system becomes a growing collection of fixes layered on top of one another. Progress slows. Certification becomes difficult. Confidence erodes.

Eventually the organization realizes that each step toward broader deployment reveals more problems than can be resolved. The system is not converging toward viability; it is chasing it. At some point, the economics no longer make sense. Funding tightens. The project stalls or collapses (Figure 2).

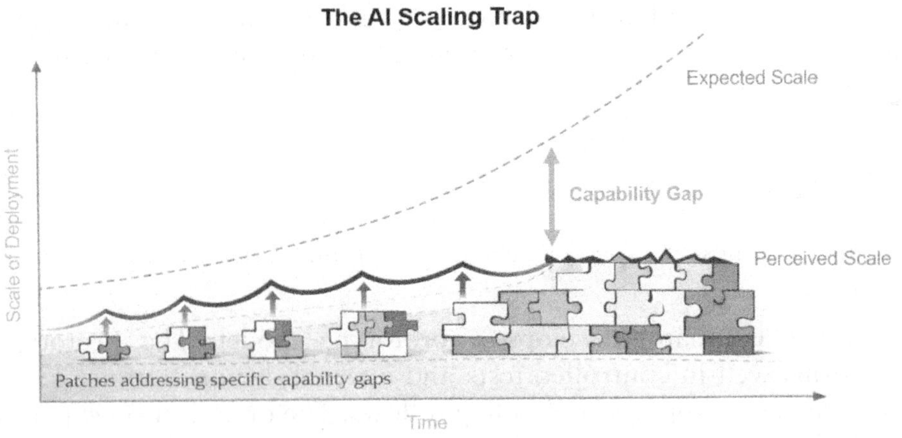

Figure 2. The AI scaling trap. Each patch temporarily raises perceived capability, but new failures emerge in deployment, so overall progress lags far behind expected scaling.

From the outside, this looks like a failure of execution or insufficient data. From the inside, it feels like running on a treadmill: constant effort with little forward motion.

What many teams miss is that they are operating under a flawed mental model of how AI systems fail. They treat deployment as an error problem: If they run enough data, they will eventually eliminate all errors. But in reality, this is not about model errors but about model behavior in context. So, will more compute fix the problem?

Scaling Is Not a Compute Problem

Today, the dominant explanation for deployment challenges is a lack of scale in training: more data, larger models, more compute. This belief has driven massive investment in infrastructure. Entire enterprise strategies are built around the assumption that larger computing capacity will automatically translate into operational success (see Figure 3 and Figure 4).

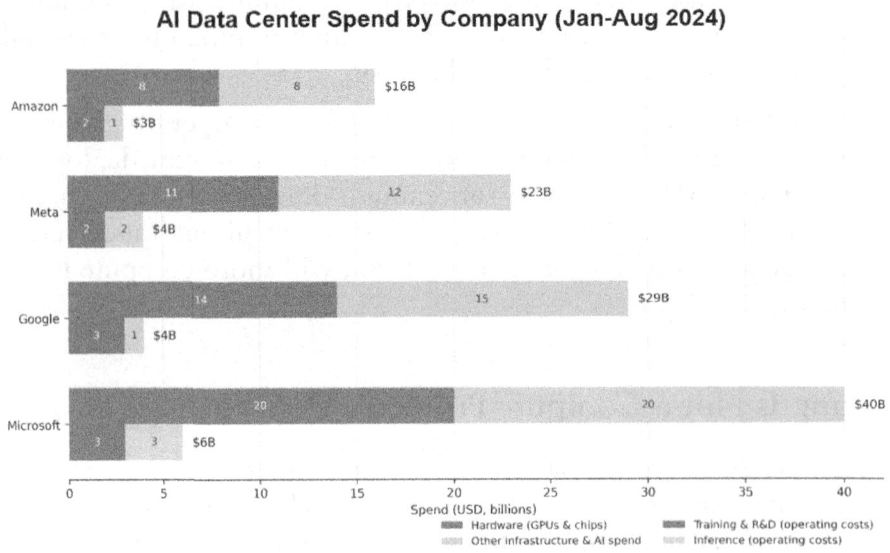

Figure 3. AI data center spend by major US companies (Data Source: Voronoi)

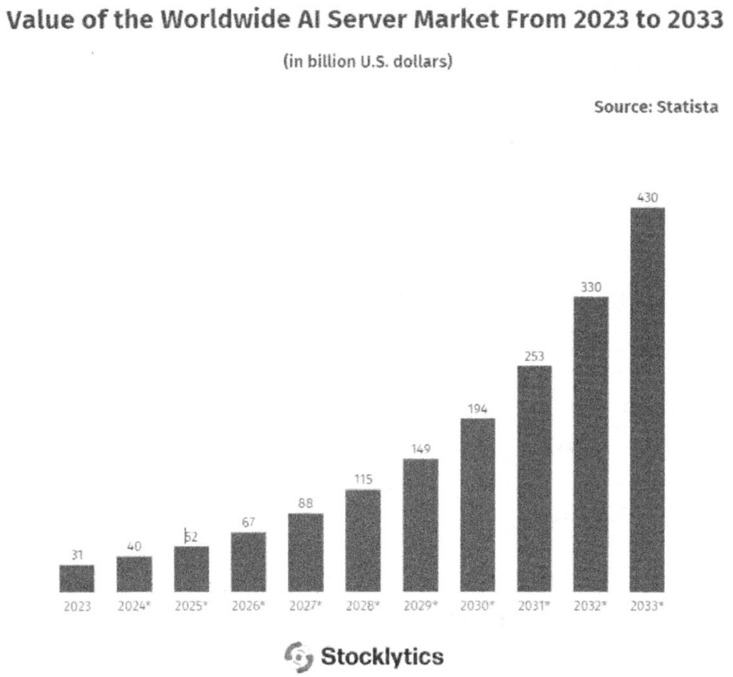

Figure 4. Growing AI server market (Source: Statista)

There is truth in this view. Larger models often exhibit broader competence. They solve more tasks and generalize better than their predecessors. But the evidence from real deployments suggests that compute alone is not the limiting factor. Specifically, organizations frequently find themselves with systems that are impressively competent in demonstrations yet cannot be trusted in production. The difference is not raw power. It is the AI's situational responsiveness.

When something does go wrong, teams often describe the issue as an "edge case"—an unexpected out-of-the-norm scenario that was not sufficiently represented in training. The implied solution is always the same: more compute to help the model generalize and eventually handle all cases, so that no new edge cases arise.

This approach is not entirely incorrect, but it doesn't scale. What we see from the present is that new edge cases continue to emerge that tend to be unaddressed by more AI training. Yes, with growing technology, the amount of compute is effectively unbounded. However, there may always be more edge cases that are different enough not to be covered by normal training. Attempting to simply upscale data processing pipelines is a losing strategy.

The deeper problem is not missing data processing capacity. It is missing the interactions that are at the heart of the continuously emerging edge cases. Teams are trying to patch failures without accounting for the interactive part that makes the difference. As a result, they are always relying on more training rather than systematically moving forward, handling more interactions, and unlocking new deployment territory. The process becomes reactive rather than proactive.

Part II—The Crux of Scalable AI

Reframing Intelligence as Behavior in Context

Part of the confusion stems from how intelligence itself is conceptualized.

In research and marketing alike, the term "intelligence" is often attributed to suggest high capability of a model: a measure of how much a model understands, how impressive the things are it can do, or often how high it scores on accepted benchmarks. These metrics can be useful as an orientation, but they obscure something essential about how systems behave outside the laboratory.

In practice, intelligence is not simply the ability to deliver peak capability. It is the ability to produce appropriate behavior in the specific situation at hand. A system that can only handle a fraction of situations and fails unpredictably in others cannot be trusted because it cannot be deployed. A system that can handle more situations, even if it means noticing its limitations and knowing what to do, can.

From a deployment perspective, intelligence is better understood as the capacity to respond appropriately across the situations it encounters.

Where "s" denotes a situation and "R" the set of real-world situations possible

This shifts the focus from peak performance to situational performance.

A model that is 99% accurate overall can still be unsafe if the remaining 1% occurs precisely in the moments that matter most (or occur most). Rare edge-case situations—not the average error—dominate real-world risk and determine whether AI is deployed successfully in critical systems like autonomous vehicles, medicine, law, aerospace etc.

Traditional benchmarks, which aggregate performance across large numbers of unrelated tasks, tend to wash out these effects. They provide snapshots of capability but little insight into behavior under the different conditions. As a result, they offer limited guidance on where and how a system can be deployed.

Scaling Is an Interaction Problem

If failures emerge from inadequate responses to situations (inadequate interactions), then evaluating models in isolation will always leave critical blind spots. For example, most current evaluation practices focus on the model alone: Given this input, what output does it produce? This approach treats intelligence as a static property rather than something that unfolds over varying conditions and sometimes dynamically over time.

But real-world deployment is inherently interactive. AI responses are good or bad depending on context—contexts in which users may adapt to systems, environments change, or incentives shift. Small variations that change context even in a small way can accumulate into large effects. Ultimately, the specific interaction at hand decides if an output is acceptable (Figure 5).

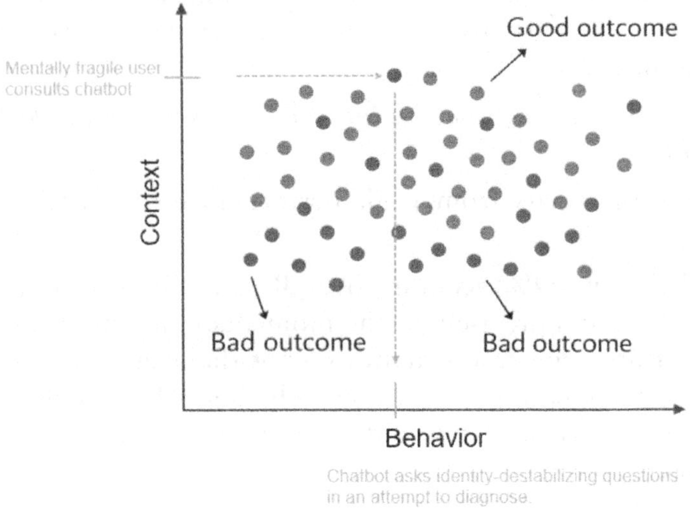

Figure 5. Describing behavior in context as interactions. Model responses produce different outcomes across contexts, so safety emerges from their combination.

Without a lens on interactions, organizations have little visibility into when their systems are robust and when they are fragile. What is missing, therefore, is not simply better models but better deployment infrastructure, mechanisms that allow teams to understand intelligence hurdles, design AI that is more prepared for their application, and succeed at deploying their AI.

In other engineering disciplines such infrastructure is taken for granted. Aircraft are not certified based solely on average performance. They are stress-tested across defined critical scenarios. Bridges are evaluated under different stress cases, not by what the bridge can do under ideal conditions. The key is to find the critical interactions and deploy systems based on evidence that they have been covered. AI systems deserve the same rigor: that the systems' intelligence covers and can handle all the cases that they may be exposed to.

Shifting the Focus: From AI Models to AI Interactions

Once this shift in perspective is made, the nature of the problem changes.

Scaling AI is no longer primarily about increasing compute and raw model capability. It becomes a question of managing interactions and enforcing adequate model behavior. Instead of asking, "What can this model do?", the more relevant question becomes "What situations in our use context can the AI handle reliably?" Deployment becomes a continuous intelligence expansion effort to progressively scale the ability to respond to situational demand.

This logic already exists in other domains. Autonomous mining vehicles operate safely within constrained environments long before similar technology is trusted in dense urban streets. The difference is not capability or model size, but the alignment between capability and situational demand. AI deployments should follow the same principle (Figure 6).

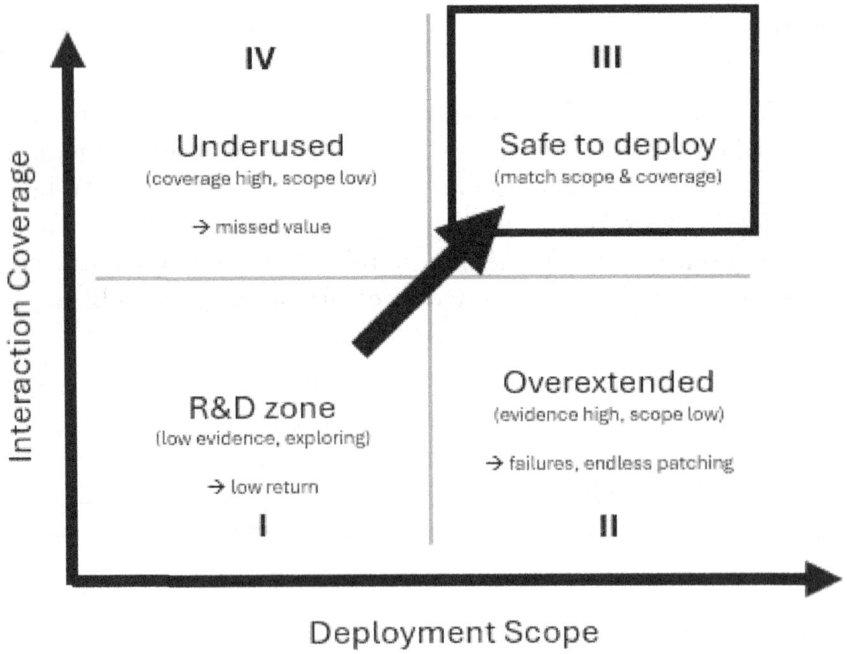

Figure 6. Interaction coverage vs. deployment scope

AI models start out in the R&D zone (area I). When scope exceeds proven capability, risk increases sharply (area II). When scope matches (area III) or is below capability (area IV), we can deploy AI with confidence, even if the model is imperfect or has errors.

The goal, therefore, is not perfection. It is targeted intelligence.

Part III—Leading the Scaling Shift

The System for Scaling AI

Translating this paradigm into practice requires a different kind of infrastructure. Organizations need ways to:

- Identify the contexts their systems will encounter
- Characterize the interactions their AI needs to master
- Determine the behavioral intelligence required in those interactions
- Detect remaining intelligence gaps as early as possible
- Close intelligence gaps before scaling up deployment

This turns scaling into a managed process rather than a leap of faith. If analysis shows that a system performs reliably in a limited set of environments, deployment begins there. If new contexts introduce unacceptable risk, the organization can choose to scale up model intelligence, add complementary safeguards, or restrict scope if desired.

Importantly, this provides options. More options. Instead of assuming that every failure must be solved through retraining, teams can make strategic decisions: Adjust the domain, redesign workflows, or do targeted interaction training. Deployment becomes a design problem, not just a data center problem. Over time, this approach produces something that is often missing in AI programs: predictability. Leaders can explain where a system can be trusted, where it can't, and why. Regulators and customers gain confidence because decisions are grounded in evidence rather than untraceable optimism. Scaling becomes deliberate rather than reactive.

Transforming Organizations for AI

For leaders, the implications are concrete. First, isolated benchmarks may no longer provide the guarantees of readiness that they are often implicated with today. They can be useful diagnostics but remain poor predictors of real-world deployment readiness.

Second, organizations must map their own operational contexts. Every deployment environment has unique characteristics—users, tool landscapes, and failure modes—that generic tests will likely not capture realistically.

Third, teams need dedicated evaluation and monitoring infrastructure. This includes people, processes, and tools designed specifically to target behavior in context, not just training bigger models.

Finally, scaling should be highly deliberate. Expansion should follow deliberately targeted intelligence, not general model capability or arbitrary scaling goals.

Taken together, these practices represent a shift away from treating AI as a data problem to treating it as an intelligence scaling problem.

Thriving in the AI Era

If the industry continues to rely primarily on larger models and broader training sets, many of today's deployment challenges will persist. The cycle of enthusiasm, overextension, and disappointment will repeat (and an AI winter may follow).

But there is an alternative path. By focusing on interactions rather than models alone, by building the required scaling infrastructure, and by scaling deliberately, organizations can close the deployment gap. The future of AI will not be determined solely by who trains the largest model. It will be determined by who learns to scale intelligence systematically and with a target. In the end, intelligence is not what AI can do. It is how well it is able to handle reality. And what ultimately determines this readiness is not raw capability but the presence of disciplined evaluation and controlled deployment processes.

About the Author

Dr. Michael Schmid is the founder of Fortify Intelligence, where he works with organizations to build and deploy AI that performs reliably in real-world settings. Over the past decade he has advised organizations including NASA, DARPA, Ford, Baidu, and defense agencies on deploying AI safely and strategically at scale.

His work focuses on the intersection of artificial intelligence, system design, and organizational transformation, with a particular emphasis on scaling intelligence through interaction-aware evaluation and targeted deployment. He contributed original language to international safety standards such as ISO 21448 (SOTIF) and developed the first risk-pattern model for AI assurance at MIT.

He advises leaders on how to move beyond model-centric approaches toward scalable AI capabilities embedded in products and operations. This chapter reflects his broader mission to help companies translate AI potential into durable real-world performance.

Email: michael.schmid@fortifyintelligence.com

Website: https://fortifyintelligence.com

LinkedIn: https://www.linkedin.com/in/drmichaelschmid/

CHAPTER 27

THE HUMAN LAYER: WHY AI SHOULD ASSIST, NOT REPLACE COMMUNICATION

By Ryan Shatzkamer
Senior Platform Engineer at outboundIQ
Boca Raton, Florida

> *Our intelligence is what makes us human,*
> *and AI is an extension of that quality.*
>
> —Yann LeCun, Chief AI Scientist at Meta

It was a little after two in the morning when the phone call came in.

A man sat on the floor of his kitchen in Budapest, his back pressed against a cabinet, the cold tile beneath him barely noticeable. His wife lay nearby, unconscious, her breathing so shallow it felt almost theoretical. Something was wrong. He did not know what. In moments like that, the name of the problem matters far less than the speed of the response.

His hands shook as he dialed the support number printed on the side of a medical device box. He spoke Hungarian.

The company on the other end of the line did not staff Hungarian-speaking agents overnight. Situations like this are where artificial intelligence earns its place without ceremony.

The call was answered immediately. The system detected the language without interruption, translated the conversation in real time, and transcribed the dialogue as it unfolded. Within seconds, the call was routed to a nurse in another country who did not speak Hungarian and did not need to. The nurse focused entirely on the patient, asking questions about breathing, responsiveness, and symptoms while the system handled translation and transcription in the background.

No one paused to admire the technology. No one announced the presence of AI. The technology stayed exactly where it belonged.

A life was stabilized because the system functioned as a bridge between people.

This is the version of artificial intelligence that makes sense.

Later that same day, the man was woken from shallow sleep by the vibration of his phone.

Then again.

Then again.

The voice on the other end of the line this time was not a nurse. It was synthetic, cheerful, and completely unaware of context. He was being informed about a financial opportunity he had not requested, could not afford to consider, and would not have answered had he been given the option to ignore it in advance.

Twelve hours earlier, technology had served as a bridge between two people in a moment of crisis. Now it functioned as a blunt instrument swung at random.

The contrast bordered on parody. The failure in this story is not artificial intelligence. The failure is intent. And intent remains stubbornly human.

Somewhere along the way, organizations decided that because AI could speak, it should speak. Because it could scale, it must scale. This logic has the comforting simplicity of a spreadsheet and roughly the same emotional intelligence.

The result has been an outbound communication ecosystem that confuses the ability to generate sound with the responsibility to communicate. Volume is mistaken for value. Automation is mistaken for improvement. The ability to reach millions of people is celebrated even when those people never wanted to be reached in the first place.

This is how communication becomes noise.

Robocalls did not become illegal because regulators are sentimental about human voices. They became illegal because they treat people as endpoints instead of participants. Automating that behavior does not modernize it. It accelerates the damage while allowing its architects to describe the process as innovation. There is something quietly absurd about calling this progress.

Artificial intelligence was supposed to make communication more intelligent. Instead, in many outbound systems, it simply made it louder. Platforms promise cost savings and headcount reduction as if trust were a rounding error that can be ignored until the quarter closes. Replacement is marketed as refinement. Speed is mistaken for relevance. Meanwhile, the person on the other end of the line is treated less like a participant in a conversation and more like a data point waiting to be processed. Trust does not respond well to this logic. Once broken, it does not rebuild itself through optimization algorithms or quarterly growth charts.

Communication, at its core, is relational. It requires attention, context, and judgment. When those elements disappear, what remains may still technically be speech, but it stops functioning as communication. Artificial intelligence did not create this problem. It simply made it easier to scale. And scale has a way of amplifying the consequences of bad decisions.

If communication begins to break down when trust disappears, the contact center is where that breakdown becomes visible first. Few environments deal with human frustration as directly as customer support. People rarely call a contact center because everything is working perfectly. They call because something failed. A payment did not process. A device stopped functioning. A flight was canceled. A medical alert triggered unexpectedly. In those moments, the conversation matters. The person calling is not looking for efficiency. They are looking for reassurance that someone understands the problem.

Artificial intelligence cannot provide that reassurance on its own. What it can do is remove the barriers that prevent humans from doing it well. In contact centers, AI does not replace the agent. It replaces friction. Used properly, AI handles the invisible tasks that consume attention and slow down communication. It transcribes calls automatically so that agents are not forced to divide their attention between listening and typing. It summarizes conversations so that context is preserved without requiring pages of manual notes. It translates languages in real time so that geography does not determine whether someone can receive help.

These functions are not glamorous. They rarely appear in marketing decks or keynote demonstrations. Yet they are the reason AI has become genuinely valuable inside well-designed communication systems. A conversation becomes easier when the person speaking can focus entirely on the other human being in the interaction.

Real-time transcription is a good example. In traditional contact centers, agents often scramble to capture details while the customer is still explaining the issue. Important information is missed. Follow-up questions repeat information that was already given. The interaction becomes slower and more frustrating for both sides.

With AI transcription running quietly in the background, the conversation becomes simpler. The system records every detail while the agent remains fully present in the moment. Instead of dividing attention between listening and documenting, the agent can concentrate on understanding the situation.

Summarization works in a similar way. When a call ends, AI systems can automatically generate structured summaries that capture the key points of the conversation. The next person who interacts with that customer begins with context rather than guesswork. Customers do not need to repeat their story multiple times across multiple interactions. This is not artificial intelligence replacing the human element. It is artificial intelligence preserving it.

Language translation provides another example of how subtle improvements create significant impact. Businesses operate globally, but language support does not always scale as easily as operations do. AI translation allows a single agent to assist customers across language barriers without delay. What once required multiple teams across different time zones can now happen in a single conversation.

The agent remains the voice of the company. The machine simply removes the linguistic barrier that would otherwise prevent the conversation from happening.

Sentiment detection adds another layer of support. These systems analyze tone, pacing, and emotional cues during a conversation. If frustration or confusion begins to rise, the system can prompt the agent with suggestions or reminders. The goal is not to control the interaction. It is to provide subtle signals that help the agent respond more effectively.

Real-time guidance works the same way. During complex interactions, agents often need to remember compliance requirements, escalation procedures, and product policies. AI systems can surface reminders at the right moment rather than forcing agents to search through documentation during the call.

Instead of interrupting the conversation to verify information, the agent receives quiet assistance exactly when it is needed. This is what well-designed AI support looks like. It is not visible to the customer. It does not impersonate the agent. It does not attempt to dominate the interaction. It simply removes the friction that prevents humans from communicating clearly.

I have worked with organizations deploying enterprise dialer systems where AI-assisted summarization and real-time guidance significantly improved outcomes. Agents began conversations with context rather than uncertainty. Patterns in customer objections became easier to identify. Compliance reminders appeared automatically when language drifted into risky territory. The difference was not dramatic in any single moment. But across thousands of conversations, the impact became obvious. Agents felt less overwhelmed. Customers felt more understood. Interactions became shorter, clearer, and more productive.

The technology did not replace the people doing the work. It helped them do it better. Problems begin when organizations reverse that relationship. When AI is used to imitate human voices and broadcast automated conversations across millions of phones, the system stops assisting communication and begins degrading it. Engagement drops. Frustration rises. Complaints accumulate. Regulators eventually intervene. None of this should be surprising.

Communication does not scale in the same way that computation does. You can multiply phone calls. You cannot multiply trust. When communication begins to scale without judgment, consequences follow quickly.

A colleague in the compliance world recently walked me through a legal complaint that bordered on the surreal. The argument was that platforms enabling automated outreach might share liability when that outreach violates consumer protection laws. The details of the case were complicated, but the underlying logic was straightforward.

When machines speak at scale, accountability does not disappear. It expands. Responsibility attaches itself to every layer of the system. The company initiating the outreach. The platform enabling it. The infrastructure routing it. The designers who built the automation in the first place. Automation does not dissolve responsibility. It multiplies it.

There is a peculiar belief circulating in parts of the technology industry that automating a violation somehow makes it less of one. A robocall delivered by a neural network remains a robocall. In some ways it is worse, because it carries the added insult of simulation. Consumers can forgive a poorly trained human. They rarely forgive being deceived by something pretending to be one. This distinction matters.

Outbound communication has always existed on a narrow ridge between relevance and intrusion. The difference between the two often comes down to timing, context, and intent. A reminder about a medical appointment is helpful. An unsolicited financial offer delivered at two in the morning is not.

Artificial intelligence did not create that boundary. It simply sharpened it.

When automation allows a company to reach millions of people instantly, the consequences of crossing that boundary grow larger. What once might have been an isolated annoyance becomes a systemic disruption. Regulators notice. Consumers file complaints. Trust erodes faster than organizations expect. Trust, once broken, does not rebuild easily.

In many industries, communication was already fragile before AI entered the picture. Customers had grown accustomed to long wait times, scripted responses, and systems that felt more concerned with efficiency than understanding. Artificial intelligence was supposed to improve that experience. In the best implementations, it does. But when AI is used primarily as a cost-cutting mechanism rather than a communication tool, the opposite occurs. Systems become louder without becoming smarter. Conversations become automated without becoming meaningful.

The technology itself is rarely the problem. Design decisions are. This is why the architecture of communication systems matters as much as the algorithms powering them. Systems must be designed with the assumption that every automated action carries consequences beyond the immediate interaction.

Compliance frameworks exist for a reason. Consumer protection laws governing outreach were not created to slow innovation. They were created because communication at scale can easily become exploitation if left unchecked. Artificial intelligence makes those guardrails even more important.

If an automated system can generate thousands of interactions per minute, mistakes can propagate just as quickly. A poorly configured campaign does not create a handful of errors. It creates thousands. The scale that makes AI powerful also makes it dangerous when misused. The solution is not to remove humans from the process. It is to design systems that assume human judgment remains essential.

Aviation offers a useful analogy. Modern aircraft rely heavily on automation. Autopilot systems manage altitude, speed, navigation, and fuel efficiency with extraordinary precision. Yet no airline advertises flights that operate without pilots. Automation reduces workload, but it does not eliminate responsibility. Pilots remain present because machines are excellent at routine execution but far less reliable when unexpected variables appear.

Communication works the same way. Artificial intelligence can analyze patterns across millions of interactions. It can surface insights, detect anomalies, and provide real-time assistance. But it cannot fully understand nuance, intention, or ethical boundaries in the way humans can. A system designed without human oversight will eventually encounter situations it was never trained to interpret. And when that happens, judgment becomes more valuable than automation.

Organizations that understand this distinction design AI systems differently. Instead of replacing human interaction, they strengthen it. They build tools that help agents understand context faster, identify risks earlier, and respond with greater clarity. The goal is not to remove the human voice from communication. The goal is to make that voice more informed.

The future of communication will not be determined by how quickly organizations deploy artificial intelligence. It will be determined by how thoughtfully they integrate it. Technology evolves faster than habits. When new tools appear, the instinct is often to maximize their reach before understanding their consequences. Artificial intelligence has followed the same pattern.

Early enthusiasm focused on scale, automation, and efficiency. Only later did the deeper questions begin to surface. What does meaningful communication look like when machines participate in the process? Where should automation stop and human judgment begin? And perhaps most importantly, what role should trust play in systems designed to operate at enormous scale?

These questions do not have simple answers. They require a shift in perspective.

For decades, contact center innovation focused primarily on speed. Calls handled faster meant lower costs and higher throughput. Automation was introduced to reduce friction in that equation. In many cases, it succeeded. Yet speed alone does not define good communication. A fast interaction that fails to solve a problem simply accelerates frustration. Artificial intelligence offers an opportunity to rethink that model. Instead of measuring success by how quickly conversations end, organizations can measure success by how clearly problems are resolved. AI systems make this possible by handling the background tasks that once consumed attention. When transcription, data retrieval, and summarization happen automatically, agents gain the freedom to focus on listening.

Listening remains one of the most underrated capabilities in modern business. Customers often call because they want confirmation that someone understands their situation. When agents are forced to juggle multiple systems while speaking, that understanding becomes harder to deliver. Artificial intelligence removes many of those distractions. Information appears in real time. Context is preserved. Guidance surfaces when necessary. The result is not faster communication. It is clearer communication.

This distinction matters more than most organizations realize. Clarity builds trust. Trust builds loyalty. Loyalty sustains businesses long after efficiency gains have faded.

Artificial intelligence cannot create trust on its own. It can, however, create the conditions that allow trust to form.

Executives often ask where AI fits best within communication systems. The answer is less dramatic than marketing narratives suggest. Artificial intelligence belongs beside the human, not in front of them. It belongs in preparation rather than impersonation. It belongs in context rather than coercion. AI remembers everything. Humans decide what matters.

AI processes information at remarkable speed. Humans decide when patience is required. AI scales effortlessly across millions of interactions. Humans build relationships one conversation at a time.

Organizations that understand this division of labor will approach AI differently. They will not see it as a replacement for people. They will see it as an amplifier for human capability. That shift in mindset changes how systems are designed. Instead of building automated outreach engines that speak endlessly, companies will build intelligence layers that help people speak more effectively. Instead of measuring success by the number of calls placed, they will measure success by the quality of the conversations that occur.

Communication has always required intention. Artificial intelligence does not remove that requirement. If anything, it intensifies it. The companies that succeed during the AI shift will not be those that automate communication the fastest. They will be the ones that protect the human layer of communication most carefully. Because technology can carry a message. Only people can give it meaning.

Artificial intelligence is not the voice of your brand. It is the amplifier. Choose carefully what you feed into it.

About the Author

Ryan Shatzkamer is a senior platform engineer specializing in enterprise contact center architecture, intelligent communication systems, and AI-assisted customer interaction. His work focuses on helping large organizations modernize both outbound and inbound communication environments through compliance-first infrastructure, advanced dialing systems, and real-time analytics platforms. Ryan has worked extensively with enterprise dialer technologies, AI-powered call analysis, and communication workflow optimization, guiding companies through the operational and regulatory challenges of scaling customer interaction in an increasingly automated world. His expertise sits at the intersection of artificial intelligence, telecommunications infrastructure, and consumer protection frameworks, where technical innovation must operate alongside legal accountability and human trust.

LinkedIn: https://www.linkedin.com/in/ryanshatzkamer/

CHAPTER 28

WHEN MACHINES DECIDE: TRUST AS THE INTERFACE OF THE AI SHIFT

By Thomas Somogyi
Head of Data, Automation, and AI in IT Consulting
Zurich, Switzerland

We have so modified our environment that we must now modify ourselves in order to exist in it.

—Norbert Wiener, *The Human Use of Human Beings*

The first time a machine makes a choice without asking, you barely notice it happening. A robot vacuum turns away from a corner it cleaned yesterday and heads somewhere new. No button was pressed; no rule was changed. It simply decided that this mattered more right now.

Most people feel a brief pause. Not fear, not excitement. Just a quiet check inside: "Is this fine? Should I stop it? Or do I let it continue?"

That brief pause marks a shift. Machines used to need instructions. You told them what to do, and they did it. Smart machines, however, act on their own assessment of a situation. They sense their surroundings, rank options, and choose a path. Their selected action might be helpful and it might even be better than your own choice.

But the unease comes from something else. The decision is no longer fully yours.

The same feeling shows up in a car that brakes before you touch the pedal. For a split second, you are no longer driving alone. The system weighed distance, speed, and risk faster than you could. You remain responsible for the outcome, yet the decision was shared.

This is the moment worth paying attention to. Not the accuracy of the system, nor its speed, but the shift in who decides.

When machines decide, what are we really trusting? The code? The data? The organization that deployed it? Or our own ability to step in when it matters?

The AI shift begins here. Not with intelligence, not with efficiency. But with confidence, reliance, and responsibility that no longer sit in one place.

Why Smart Machines Feel Different from Traditional Technology

People learned to trust machines through repetition. A device behaved the same way yesterday, so it would behave the same way today. When something failed, the reason usually sat close to the surface. A switch broke. A sensor misread. The rules stayed stable, even when the outcome did not.

Smart machines interrupt that rhythm. They still follow constraints, yet their behavior shifts with context. The system notices small differences, weighs them, and acts on what seems most appropriate at that moment.

This is where the tension emerges. Humans rely on habits formed around consistency; we expect machines to behave correctly in the same way each time. Adaptive systems, however, aim for the correct outcome regardless of the steps. Both approaches are reasonable, but they rest on different assumptions.

The difference becomes harder to ignore once machines move through the physical world. Software that suggests a movie or flags an email stays at a distance. A machine that moves, steers, or stops does not. Its choices unfold in real time, in a shared space, where hesitation and correction carry weight. Even when outcomes improve, the experience feels less forgiving.

What unsettles people is not loss of control in a technical sense. It is loss of predictability in a familiar form. We struggle with reasonable unpredictability, even when it is safer on average.

Seen this way, resistance has little to do with capability. It grows from a mismatch between long-standing expectations and a new kind of behavior. Smart machines ask humans to adjust how they judge reliability. That adjustment is psychological first, technical second.

Trust as the True Interface of the AI Shift

For a long time, interfaces were easy to spot. Buttons, screens, switches, and menus made it clear where control lived. You pressed something, the system responded, and the relationship stayed visible. If the outcome felt wrong, you knew where to point your attention.

Smart machines change that arrangement. The most important interface is no longer visible. It now sits in the space between a human and a system that can act on its own judgment. Confidence replaces clicks. Judgment replaces instruction. The interaction begins before any screen appears.

Trust plays a different role in this setting. It is not belief in technology, and it is not optimism about progress. Trust functions as operational confidence. It answers a practical question: Can I rely on this system in this situation, and do I know when to step in? That answer shifts with experience and context. A system trusted in one setting may feel wrong in another, even if nothing technical has changed.

This dynamic explains two common failure patterns. Under-trust leads to constant overrides. People ignore or disable systems that might help because the cost of surprise feels too high. Over-trust moves in the opposite direction. The system works well often enough that vigilance fades, and when something unusual happens, the human responds too late.

Both patterns grow from the same mistake. Trust gets treated as a feature that appears once a system performs well enough. In reality, trust is shaped by design, communication, and responsibility. It can rise or fall without affecting accuracy.

Seen this way, trust is the real interface of the AI shift. It determines how autonomy gets used, limited, or rejected. Trust does not need to be left to chance. It can be designed. Doing so requires breaking trust into recognizable parts.

The Five Dimensions of Trust in Autonomous Systems

Trust sounds abstract until it breaks. People usually notice it only after something feels wrong, when confidence drops or hesitation sets in. To make trust workable, it helps to break it down into parts that people recognize from experience.

One practical way to think about trust in autonomous systems is through five dimensions: safety, predictability, transparency, control, and accountability. Each addresses a different question people ask, often without realizing it. Together, they explain why a system that performs well can still feel wrong and why another earns confidence even when it is imperfect.

Safety

Safety is the starting point for trust. Before people think about usefulness or performance, they ask a simpler question: Does this system feel safe to be around?

Safety operates on several levels at once. Physical safety concerns harm to bodies and property. Operational safety reflects how a system behaves during normal use and edge cases. Systemic safety shapes how failures spread, stop, or escalate. All three matter, though they do not carry equal weight in human judgment.

Actual safety and perceived safety often diverge. A system can meet every internal target and still fail to gain acceptance. People do not experience safety through metrics alone; they experience it through moments that stick. One unexpected movement or unexplained action can outweigh months of reliable behavior.

This explains why early resistance appears even when records look strong. A machine that is safer than a human alternative may still feel unsettling. People judge risk through lived experience, asking whether they can anticipate what might happen next and whether they would have time to respond if something goes wrong.

Physical systems amplify this effect. When autonomy moves through shared space, hesitation carries weight. The cost of being wrong feels immediate, and people respond by keeping distance, intervening early, or refusing reliance altogether.

Designing for safety, then, means more than reducing failure rates. It means shaping how safety is communicated, how boundaries are signaled, and how recovery looks when something unexpected occurs. Trust grows when harm feels unlikely, and surprises do not spiral out of control.

Predictability

Predictability does not mean identical outcomes. People do not expect machines to behave the same way every time. What they look for instead are boundaries they can rely on, a sense of what a system will not do even when conditions change.

Trust grows when that logic becomes legible through experience. Humans tolerate variation when patterns are visible over time. A system that reacts differently in different situations can still feel predictable if its responses make sense after the fact. Confusion appears when actions feel disconnected from what people observe or expect.

This is where many systems struggle. Predictability gets confused with repeatability. Designers focus on performance, while users look for consistency in behavior. When a system surprises people without warning or framing, confidence erodes, even if the result itself is sound.

Predictability works at the level of expectation management. Clear behavioral limits and recognizable escalation patterns matter more than precision. Without them, people intervene early or avoid reliance. With them, uncertainty becomes livable.

Transparency

Transparency is often misunderstood. It is not about exposing internal logic or walking people through how a system arrives at a result. What they want instead is clarity about intent, limits, and what might happen next.

A transparent system helps people understand why it acted the way it did. It signals where confidence ends and uncertainty begins. It makes clear what the system can handle on its own and where human judgment is expected to step in.

Problems arise when systems stay silent. A decision happens without context, leaving people to guess whether the action was deliberate, cautious, or accidental. When intent is hidden, even reasonable behavior can feel arbitrary. Over time, this pushes people toward early intervention or avoidance.

Transparency works when systems communicate in human terms. A visible boundary, a simple signal of intent, or a clear indication that conditions have shifted is often enough.

Many so-called black box complaints point to this missing layer. They are less about hidden logic and more about absent signals. When intent and limits are clear, autonomy becomes easier to accept, even when outcomes vary.

Control

Control shapes how comfortable people feel around autonomous systems. It is less about frequent intervention and more about knowing that intervention remains possible. The option to pause, override, or disengage changes how people relate to autonomy, even when it is never used.

Systems that remove control entirely tend to raise tension. Users stay alert, second-guess decisions, and hesitate to rely on the system at all. The absence of an exit increases the perceived cost of every action. By contrast, systems that preserve a clear way out invite calmer use. Optionality lowers anxiety long before it is exercised.

Control works best as reassurance, not supervision. Its presence signals respect for human judgment. It reminds people that they remain part of the decision loop, even when the system acts on its own.

Problems arise when control is unclear. If people are unsure when they can intervene, or whether intervention will be honored, confidence drops quickly. In response, they become overly cautious or avoid the system altogether.

Designing for control means making boundaries visible and dependable. Autonomy becomes easier to accept when people know they can step in without friction. Trust grows not from constant oversight, but from the quiet assurance that control remains within reach.

Accountability

Accountability becomes visible when outcomes are uncertain. As long as systems behave as expected, responsibility stays in the background.

When something goes wrong, people look for a human answer. Automation does not remove that need. It makes it more explicit.

A common breakdown appears when responsibility dissolves into the system itself. Phrases like "the system decided" sound neutral, yet they quietly avoid ownership. Decisions may be delegated, but responsibility still needs a place to land. When no one clearly owns the outcome, trust erodes quickly.

This is where the difference between fault and responsibility matters. Fault asks who caused an error. Responsibility asks who defines intent, sets boundaries, and responds when conditions change. In autonomous settings, these roles must be clear before autonomy is granted. Waiting until after something goes wrong is already too late.

Accountability does not require perfect foresight or technical certainty. It requires clarity about who monitors behavior, who can intervene, and who stands behind the result when judgment is shared. When these roles are visible, mistakes remain manageable. When they are blurred, even small issues can feel unsettling.

Trust survives errors. It rarely survives evasion. People are more willing to rely on imperfect systems when accountability remains stable and human. Autonomy works best when responsibility does not disappear, even as decisions become distributed.

When Trust Breaks: Common Failure Patterns

Trust does not fail in a single way. Sometimes it erodes through small, repeated frictions. Other times it collapses in a single moment. In both cases, the break tends to appear where humans and systems meet. The technology may perform as expected, yet the interaction fails.

One common pattern emerges when automation runs without clear escalation paths. Systems act confidently until they reach conditions they were never meant to handle. When that happens, there is no visible handoff and no clear signal that human judgment is needed. Actions occur without context or intent, and from the outside, the behavior feels arbitrary, even when it follows internal logic. People notice only after outcomes begin to feel wrong.

Responsibility gaps deepen the problem. Decisions spread across systems, teams, and processes, while ownership remains vague. When something goes wrong, responsibility drifts. The final break often comes when humans are blamed for outcomes they could not influence. Intervention was unclear or impossible, yet accountability lands on the operator. That mismatch damages trust faster than most technical errors.

Designing for Trust: Practical Patterns for Humans and Organizations

Trust does not appear after deployment. It is shaped long before a system goes live, through deliberate choices about roles, limits, and responsibility. Organizations that treat trust as a technical outcome often notice the gap too late.

The first pattern is clarity at the boundary. Humans and machines should not negotiate decisions in real time. Clear lines define which decisions stay human, which can be delegated, and under what conditions control shifts. Boundaries reduce fear because they replace guesswork with expectation.

Handover points matter just as much. Systems should signal when conditions change and when human attention is required. When escalation paths are visible, people respond earlier and with greater confidence.

Another pattern is designing for graceful degradation. No system performs well in all conditions. Trust grows when systems slow down, limit scope, or hand back control as uncertainty rises.

Responsibility ties these patterns together. Accountability should be set up front, not negotiated after an incident. Trust is governed before it is earned. When intent, ownership, and response are clear, autonomy becomes workable rather than risky. Seen this way, trust is a design choice that shapes how people and technology grow together.

Thriving Alongside Smart Machines: A Durable Mindset

Thriving alongside smart machines begins with accepting uncertainty as a permanent condition. Autonomy does not remove ambiguity; it redistributes it. Systems act faster and across wider contexts, yet the need for human judgment does not disappear. It simply takes a different form.

This reframes the human role over time. People are no longer defined by constant execution, but by supervision and sense-making. Humans remain responsible for setting intent, recognizing when conditions shift, and deciding when autonomy should slow down or step back. Machines can support decisions, but they do not own their consequences.

Trust in this setting is not a milestone that gets reached and left behind. It evolves through use, correction, and occasional friction. It grows when systems behave within understood limits and when people know how to respond when they do not. It weakens when responsibility becomes unclear or when judgment fades quietly into automation.

Thriving does not mean surrendering judgment to technology. It means designing systems where human responsibility remains intact, even as autonomy expands. Clear roles, visible boundaries, and explicit accountability make that possible.

There is no need for optimism or fear here. Smart machines will continue to shape how decisions are made. The lasting advantage lies in shaping how trust, judgment, and responsibility move with them.

About the Author

Thomas Somogyi is a business and IT strategist working at the intersection of artificial intelligence and human decision-making. With nearly 20 years of experience across software development, consulting, and leadership, he brings a structured, system-oriented perspective to digital transformation. He holds an EMBA in business psychology, which shapes his focus on how technology influences human behavior and organizational responsibility.

Drawing on experience ranging from service in the Swiss Army Special Forces to leading data and AI initiatives in complex enterprise environments, Thomas advocates for technology that serves human intent rather than replaces it. His work emphasizes that while machines can calculate and optimize, accountability and judgment remain fundamentally human disciplines.

Through a system-thinking lens, he helps organizations navigate the shift toward autonomy with clarity, structure, and accountable governance, ensuring that innovation strengthens resilience instead of eroding trust.

E-Mail: info@thomassomogyi.com

Website: www.thomassomogyi.com

CHAPTER 29

THE RISE OF MICROENTREPRENEURS AND AIPOWERED SOLOPRENEURS

By Sakina Syed
Microsoft IT Professional, AI Engineer and Consultant
Toronto, Canada

The best way to predict the future is to invent it.

—Alan Kay

For most of modern history, building a business required an army. You needed a marketing team, a finance department, a designer, a copywriter, a customer service desk, and—if you were lucky—a visionary leader who could keep all the moving parts aligned. Entrepreneurship was a highrisk, highoverhead endeavor, accessible only to those with capital, connections, or the willingness to sacrifice years of stability.

But something extraordinary is happening. A quiet revolution is unfolding, one that is reshaping the very definition of what it means to be an entrepreneur. Today, a single individual with a laptop and an internet connection can build a brand, launch a product, reach a global audience, and scale a business faster than entire companies could just a decade ago. This is because they now have access to something that functions like a superpower: artificial intelligence.

AI is becoming the great equalizer. It's dissolving barriers that once kept people from pursuing their ideas. It's amplifying creativity, accelerating execution, and giving individuals the ability to operate with the efficiency of a full team. The solopreneur of the past was often overwhelmed, juggling tasks late into the night. The solopreneur of today is augmented due to their support by intelligent tools that handle the heavy lifting, leaving humans free to focus on vision, strategy, and connection.

This is not a story about machines replacing people. It's a story about people finally being able to do the work they were meant to do without being buried under the weight of everything else.

We are entering the age of the AIpowered microentrepreneur: the oneperson powerhouse capable of building something meaningful, profitable, and scalable. And this shift is not just technological. It's cultural. It's economic. It's deeply human.

The rise of AIenabled solopreneurs is one of the most exciting transformations of our time, and it's only just beginning.

Why This Moment Matters—The Perfect Storm of Opportunity

Revolutions rarely happen because of a single breakthrough. They happen when multiple forces converge at the same time, creating a tipping point. That's exactly what we're witnessing now.

First, AI tools have become accessible. You no longer need a PhD in machine learning to use them. They're intuitive, conversational, and increasingly integrated into everyday platforms. What used to require specialized knowledge is now available to anyone with curiosity and a willingness to experiment.

Second, remote work has normalized independence. Millions of people discovered during the pandemic that they could be productive outside traditional office structures. Rather than fade, that shift evolved. People now crave autonomy, flexibility, and meaningful work. Entrepreneurship no longer feels like a radical leap; it feels like a natural extension of how we already operate.

Third, digital marketplaces have exploded. Platforms like Etsy, Shopify, Gumroad, Patreon, and Substack have made it possible to sell products, services, and content directly to global audiences. Distribution is no longer the bottleneck. The world is your market.

Fourth, the cost of starting a business has plummeted. What once required thousands of dollars in software, design, and labor can now be done with free or lowcost AI tools. The financial risk is lower than ever, which means more people are willing to try.

And finally, there's a psychological shift underway. People are realizing that entrepreneurship is not reserved for the elite. It's becoming democratized. The question is no longer "Can I start something?" but "What do I want to create?"

This perfect storm—accessible AI, cultural change, global distribution, low startup costs, and rising confidence—has created a onceinageneration opportunity. The door to entrepreneurship has swung wide open, and millions are stepping through. Once inside, a new reality greets them: one person can now do the work of many.

The AI Toolbelt—How One Person Can Now Do the Work of Many

To understand the rise of AIpowered solopreneurs, you have to understand the "virtual team" AI provides. These tools do more than automate tasks. They replicate entire roles that once required specialized skills.

AI as Strategist

AI can analyze markets, identify trends, and validate ideas with astonishing speed. It reveals what customers are searching for, how competitors are positioning themselves, and where untapped opportunities exist, acting like a research department that works instantly and never sleeps. For example, AI can highlight emerging keyword trends, forecast which product categories are gaining momentum, map gaps in competitor offerings, and pinpoint the challenges customers voice most often. These are all insights that once required entire teams weeks to uncover. AI Tools include Perplexity, which delivers realtime market and competitor intelligence, and Aomni, which compiles comprehensive AIdriven industry reports, giving solopreneurs immediate clarity on trends, opportunities, and strategic direction. Strategy is only the beginning. Once you know where to play, AI helps you create what to ship.

AI as Creative Team

Branding, design, copywriting, video editing—tasks that once required hiring multiple professionals—can now be done with AI tools that generate logos, write marketing copy, create social media content, and produce videos. The quality is high, the turnaround is immediate, and the cost is minimal. Tools like Adobe Firefly, which creates brand-safe commercial visuals and editable vector assets, and Midjourney, known for producing high-quality concept art and stylized imagery, significantly accelerate creative workflows. Platforms such as Canva's AI Suite allow anyone to generate graphics, layouts, and social content in seconds, while Runway and Pika offer text-to-video generation and AI-powered editing, bringing professional-level video creation within reach of solopreneurs. Beyond creation, AI keeps the business humming by handling the operational backbone.

AI as Operations Manager

Behind every thriving solo business is an invisible engine keeping the gears turning, and today, that engine is AI. From scheduling and invoicing to workflow automation and email management, AI takes over the administrative grind that once drained time and energy. In fact, businesses using AIdriven workflow automation report saving up to 30% of the hours previously spent on manual operational tasks, freeing solopreneurs to reinvest that time into strategy, revenue generation, and creative work. There are awesome tools such as GenFuse AI that act as a fullscale operations command center, automating multistep workflows, connecting seamlessly with apps like Gmail, Google Calendar, Slack, and Notion, and transforming naturallanguage instructions into sophisticated automations that work around the clock. And for entrepreneurs rooted in the Microsoft ecosystem, Outlookintegrated assistants like Lindy supercharge daily productivity by triaging email, coordinating schedules, and handling followups directly inside Outlook, bringing an intelligent, everpresent operations partner into the heart of the Microsoft 365 workflow. With the back office on autopilot, AI can also become the front line for customer experience.

AI as Customer Support

Chatbots and AIpowered email assistants can respond to inquiries, troubleshoot issues, and maintain customer relationships. They provide consistency, speed, and availability that would be impossible for one person to manage alone. Tools such as Intercom and Ultimate.ai automate routine customer questions and sync with CRM data to deliver rapid, accurate responses, giving solopreneurs enterprise-level support without hiring a team. While customers are cared for, AI can simultaneously distill the data behind the scenes into decisions.

AI as an Analyst

Data is the lifeblood of modern business, but interpreting it can be overwhelming. AI can analyze performance metrics, forecast trends, and recommend optimizations. It turns raw information into actionable insight. Agentic analytics platforms such as Anomaly AI can autonomously inspect data, clean anomalies, generate insights, and build dashboards, acting like a full analytical team that works behind the scenes. The Analyst agent in Microsoft 365 Copilot functions like a virtual data scientist, analyzing spreadsheets, identifying trends, running Pythonpowered insights, and transforming raw data into clear reports and forecasts, all within the Microsoft 365 ecosystem.

Together, these tools form a digital workforce that empowers individuals to operate with the efficiency of a full team. The solopreneur is no longer alone. They're supported by an intelligent ecosystem that amplifies their capabilities.

The Rise of the MicroBusiness Model

So what exactly is a microentrepreneur in the age of AI?

It's someone who builds a business that is intentionally small, agile, and scalable. They don't aspire to hire dozens of employees or build a corporate empire. Instead, they leverage AI to create a lean, efficient operation that supports their lifestyle, values, and goals.

Emerging archetypes include:

- *AIPowered Consultants*: experts who enhance delivery with AI, offering faster insights and more client capacity without burnout
- *OnePerson Digital Agencies*: solo providers delivering branding, marketing, or content at enterprise quality via AIaccelerated production
- *Creators with AI Pipelines*: writers, podcasters, YouTubers, and educators using AI for ideation, editing, repurposing, and distribution

- *Niche Product Builders*: entrepreneurs crafting templates, courses, apps, and guides with AIassisted design, testing, and refinement
- *Educators & Course Designers*: subjectmatter experts structuring curricula, generating visuals, and producing highquality learning experiences with AI

These models share a common thread: They scale without traditional headcount. Growth comes from leverage, not labor. AI becomes the multiplier that allows one person to achieve what once required a team. And the best evidence that this isn't theoretical? The people already living it.

RealWorld Transformations—Stories of AIEnabled Solopreneurs

The most compelling evidence of this shift comes from the people living it.

Case Study 1: The Designer Who Became a Studio

A freelance graphic designer used to spend hours creating concepts for clients. With AI tools, she now generates multiple design directions in minutes, refining them with her artistic eye. Her output has doubled, her rates have increased, and she now serves clients around the world without hiring a single employee.

Case Study 2: The Educator Who Built a Global Classroom

A former teacher wanted to create online courses but felt overwhelmed by the production demands. AI helped her script lessons, design visuals, and edit videos. Within six months, she built a thriving digital academy with thousands of students.

Case Study 3: The Consultant Who Scaled Without Burnout

A business strategist used AI to automate research, generate reports, and analyze client data. What once took days now takes hours. He serves more clients, earns more income, and has reclaimed his evenings and weekends.

These stories are not outliers. They represent a growing movement of everyday people using AI to unlock their potential. The common thread is empowerment. AI doesn't replace their talent; it amplifies it.

Case Study 4: The Small Retailer Who Became a One-Person Commerce Engine

A boutique home-goods retailer used to spend hours each week managing inventory, updating product listings, writing descriptions, and responding to customer inquiries. With AI tools, she now automates her product photography edits, generates polished descriptions and social media posts, forecasts inventory needs, and handles customer questions through an intelligent chatbot. What once required a part-time assistant and late nights now happens automatically in the background. Sales have grown, return rates have dropped thanks to clearer product details, and she's expanded from a local shop to a global online storefront, all while remaining a business of one.

Case Study 5: The Retired Professional Who Turned Experience Into an AIPowered Income Stream

A retired financial advisor spent decades building expertise but worried that stepping away from fulltime work meant losing both relevance and income. Instead of starting over or taking on demanding consulting hours, he began using AI to transform his lifelong knowledge into automated value. AI tools helped him convert his signature workshop into a polished online course, generate ongoing email newsletters, and even create a personalized financialwellness chatbot that answers common questions his past clients used to ask. What once took him days of writing, editing, and prep now takes minutes. The best part? His content continues to earn passive income each month while he enjoys retirement, proving that AI doesn't just empower new entrepreneurs, but gives seasoned professionals a way to keep contributing, keep earning, and keep their legacy alive without the grind.

The Economic and Social Impact of the Solopreneur Revolution

The rise of AIpowered microbusinesses is reshaping the economy in profound ways.

Local and Global Economic Growth

Microentrepreneurs contribute to economic resilience. They create new markets, serve niche audiences, and stimulate local economies without requiring large infrastructure.

The Decline of the Corporate Ladder

Traditional career paths are giving way to "microempires": small, independent ventures that offer autonomy, creativity, and financial stability.

More Innovation, More Voices

When barriers fall, more people participate. This leads to a richer, more diverse landscape of ideas, products, and solutions.

Experience, Amplified: AI's Economic Boost for Aging Populations

AI is also reshaping work for aging populations, allowing retirees to continue contributing on their own terms. By transforming decades of expertise into digital products, courses, and automated services, seniors can maintain income, stay mentally engaged, and participate in the economy without the demands of traditional employment.

Opportunity Creation Over Job Seeking

Instead of waiting for companies to hire them, people are creating their own opportunities. This shift has the potential to reduce unemployment, increase job satisfaction, and foster a culture of selfdetermination.

The solopreneur revolution is a social transformation. And underpinning that transformation is a change in identity and intent.

The Mindset Shift—From Employee to Creator

Perhaps the most profound change AI enables is psychological. For decades, people were conditioned to think of themselves as employees: cogs in a larger machine. Stability came from fitting into predefined roles, and creativity was often secondary to compliance.

AI is flipping that script. It encourages experimentation, rewards curiosity, and lowers the cost of failure, making it safer than ever to pursue new ideas. It reframes stability, not as holding a secure job, but as possessing the ability to create value in multiple ways.

And at the heart of this shift lies a deeper truth: In the industrial age, power belonged to those who owned the means of production; in the intelligence age, it belongs to those who own the means of creation. AI places those means directly into human hands.

This transition and mindset shift—from dependency to autonomy—is one of the most empowering aspects of the AI revolution. People are rediscovering their agency. They're realizing they can build, create, and innovate on their own terms. Work is no longer simply a place you go or a role you're assigned. It becomes a capability you cultivate and a portfolio of value you compose.

Challenges and Ethical Considerations

No transformation is without challenges, and the rise of AIpowered solopreneurs is no exception. But these challenges are solvable.

Overreliance on Tools

AI should augment human creativity, not replace it. The key is balance: using AI as a partner, not a crutch.

Skill Gaps

Some people may feel intimidated by new tools. The solution is education, mentorship, and accessible training.

Ethical Use of AI

Transparency, fairness, and responsible use are essential. Solopreneurs must ensure their work aligns with ethical standards.

Market Saturation Myths

People often fear that "everyone is doing it." But markets expand with creativity. Differentiation comes from authenticity, not volume. Your voice, your story, and your unique way of solving problems will always set you apart, and there is room for you. Don't let the illusion of saturation stop you from starting. Most people never take the first step, which means your willingness to begin is already a competitive advantage.

These challenges are real, but they are not roadblocks. They are simply part of the learning curve of a new era. And that curve is bending toward opportunity.

The Future—A World Where Millions Build Their Own Path

Imagine a world where millions of people wake up each morning not to clock in, but to create. A world where the first question of the day is not "What do I have to do?" but "What do I want to build?" Imagine creativity no longer treated as a rare talent or a weekend hobby, but as a basic human instinct, supported, amplified, and brought to life by intelligent tools that make innovation effortless.

This world is unfolding right now, and we are watching the foundations being laid in real time. In the next decade, we will see a profound reshaping of how individuals participate in the global economy:

- AInative entrepreneurship woven into education, where students learn to prototype businesses before they can drive, using AI as tutor, researcher, and production partner.
- Personalized AI "cofounders" that know your strengths, fill your gaps, and help you transform ideas into companies, drafting plans, validating markets, and building assets alongside you.
- Global networks of solopreneurs, collaborating fluidly across borders, time zones, and languages, forming dynamic microteams that assemble and dissolve around opportunities.
- Vibrant ecosystems of microbusinesses that collectively rival traditional corporations in creativity, speed, and impact, yet remain lean, humancentered, and deeply aligned with personal values.

This rise of AIpowered solopreneurs is far more than a technological shift. It is a cultural reawakening. A reclaiming of agency. A redefinition of who gets to innovate, who gets to build, and who gets to participate in shaping the future.

For the first time in history, entrepreneurship is no longer gated by capital, connections, or geography. The tools once reserved for large companies now sit on the desks and in the pockets of everyday creators.

AI is not taking over the world. It is helping people take ownership of theirs. We are entering an era where individuals can scale ideas with the leverage of entire teams, where creativity meets capability without friction, and where the courage to try is finally matched by the tools to succeed.

And the most exciting part? We are only at the beginning. The next chapter of entrepreneurship won't be written by institutions. It will be written by individuals—millions of them—empowered by AI, united by possibility, and driven by the timeless human desire to create something meaningful.

About the Author

Sakina Syed is a dynamic senior data and AI consultant, AI engineer, and enthusiast with a stellar track record in the computer software industry, including notable tenures at tech giants like Microsoft. As a Microsoft AI engineer and Azure-certified professional, she excels in AI deployment, sales, management, teamwork, and leadership.

With a Bachelor of Science in Neuroscience and Mental Health Studies from the University of Toronto, Sakina has also enriched her expertise with courses in artificial intelligence, business management, and IT. Her passion for writing and traveling adds a unique dimension to her professional persona.

Sakina's enthusiasm for business and the application of business psychology to understand consumer needs is matched by her extensive seven-year experience in customer service. She is adept at interpreting consumer feedback and statistical data, enabling her to identify market requirements with precision. Her strong passion for team management and leadership drives her dedication to delivering exceptional customer experiences and fostering business success.

Email: sakina@youraiconsultant.ca

Website: www.youraiconsultant.ca

LinkedIn: https://www.linkedin.com/in/sakina-syed/

CHAPTER 30

INVESTING IN INTELLIGENCE: WHY MONEY, WORK, AND GOVERNANCE WILL BE REWRITTEN

By Alex Szomora
Tech investor and Go-to-Market strategist
Singapore

The challenge with AI is not just building it—it's making sure it aligns with human values.

—Sam Altman

If you put AI builders, safety researchers, economists, and philosophers in the same room, you'd expect chaos. Instead, you get an oddly consistent chorus: AI is getting powerful fast, but the limiting reagent isn't intelligence, it's institutions. The frontier crowd is quietly admitting that "just scale it" isn't a strategy anymore. The next gains come from post-training, rigorous evaluation, multimodality, and embodiment while reliability stays moody: brilliant one minute, confidently wrong the next. Governance folks are basically waving a neon sign that says: "If oversight, evaluations, and compute-level controls don't scale with capability, deployment will outrun trust." Meanwhile, economists are doing what economists do: reminding everyone that diffusion is slow, organizations are messy, energy and capital concentrate, and productivity gains will not magically trickle down unless AI is designed to complement people rather than treat them like a legacy system. Philosophers add the final punchline: Legitimacy isn't a model metric. It's context, dignity, and responsibility. Privacy is about norms, not just data; extraction without reciprocity triggers backlash; and in regulated, high-stakes domains like health, finance, and biosecurity, it is the rarest commodity on the shelf. The dominant trend isn't "race to autonomous superintelligence." It's a shift toward AI as a framework where evaluation, governance, incentives, and institutional design determine whether AI amplifies human flourishing or just industrializes fragility.

My stance is neither utopian nor dystopian. AI isn't a force of nature; it's an accelerator of human incentives, like giving a megaphone to whatever your system already rewards. The real disruption won't be "machines replacing humans" so much as intelligence deployed faster than governance, capital discipline, and social contracts can adapt. You get chaos not from evil AI, but from very competent automation colliding with very human incentives.

Capital markets already see the plot. PitchBook shows AI/ML funding is enormous and increasingly shaped by big rounds and infrastructure-heavy plays. McKinsey's tech trends reporting points to the same concentration dynamics across major themes. a16z documents enterprise GenAI graduating from innovation budgets into durable software line items, meaning the CFO has entered the chat, and now everything is about ROI. As adoption matures, novelty is being replaced by a very unsexy "show me the money" phase: Deployment friction, reliability, integration, and org-change are now the binding constraints. The winning products will look less like chat UX and more like infrastructure tucked into workflows. Agents are the headline, but reliability and measurement are the bottlenecks, so "agent ops," evaluation, monitoring, and audit tooling are suddenly hot. And as Satya Nadella keeps reminding us, treat AI like a new productivity layer across roles, but ship it "safe and secure from the outset," because nothing kills a rollout like a compliance team discovering your agent has opinions. Think of AI not as a new type of vehicle, but as a new high-speed rail system. Investing in intelligence isn't just about buying the fastest train out there (the model), it's about owning the tracks, the stations, and the ticketing systems (the infrastructure). If the tracks don't go to every neighborhood (inclusion) or if people don't trust the safety of the brakes (governance), the train, no matter how fast, will never reach its destination.

A New Economic Stack Is Emerging

We're watching AI assemble a new economic stack, almost like a startup pitch deck, except it's actually happening. Capital is concentrating into a small number of models. Regulation is fragmenting into regional rulebooks, turning compliance into a moat (and occasionally a migraine). And money is becoming programmable—stablecoins and tokenized cash—so AI can transact, settle, and enforce policy at machine speed.

In that world, the biggest opportunities aren't "yet another AI app." They're trust rails: systems that connect AI, identity, payments, auditability, and cross-border compliance so AI can operate safely in finance, health, and critical architecture. That's the bet: The next phase of AI adoption will reward people who treat AI as economic and institutional architecture, not just software. Until then, AI remains—aptly put on Manifold—a "brilliant but unreliable genius colleague," capable of very simple and very profound mistakes. (If you've ever hired a genius intern who sometimes sets the kitchen on fire, you get it.)

Who Benefits, Who Is Harmed

AI won't distribute upside politely. It will disproportionately benefit regulated incumbents and compliance-native challengers that already understand risk, liability, and audit trails. It will favor ecosystems that invest in data fidelity and interoperability, where AI can be evaluated, traced, and monitored, not just shipped with vibes. And it will reward operators—individuals and teams—who use AI as leverage by embedding judgment, workflows, and verification into daily practice.

On the flip side, it will disproportionately harm workers without reskilling pathways or bargaining power, where productivity gains don't translate into shared upside. It will punish markets that allow extraction without reciprocity, where data and work are captured but value isn't returned. And it will burn companies that optimize for speed without accountability, especially in high-stakes domains like health, finance, and security, where "oops" is not a product feature.

AI Capital Is Concentrating, Not Democratizing

There was once a thesis that AI would democratize opportunity. In practice, the architecture requires deep capital, talent wars are real, and rich ecosystems get richer. Democratization will happen mostly at the edges where regulatory and domain complexity create barriers that big players find too annoying or too expensive to navigate. Meanwhile, VC and private funding is pouring into AI, but a huge share is concentrating into mega-winners and ecosystems (with heavy corporate participation), creating a barbell market: a few dominant platforms and a long tail of fragile application-layer startups.

Hugging Face illustrates what "democratization at the edges" means in practice. While frontier model development and compute are consolidating into a small number of capital-intensive winners, Hugging Face built leverage by owning the layer big players find inconvenient: openness, evaluation, governance, and interoperability. Rather than competing on raw model scale, it became the default infrastructure for model sharing, benchmarking, fine-tuning, and increasingly compliance-friendly deployment, particularly relevant in regions like the EU where transparency, auditability, and reproducibility are not optional. Its advantage isn't superior models, but institutional fit: making it possible for enterprises, researchers, and regulators to inspect, compare, and govern AI systems. In that sense, Hugging Face doesn't contradict concentration. It complements it. The core concentrates, the edges democratize through tooling that turns regulatory and governance complexity into a defensible moat.

You can see the money moving toward safety and governance tooling, AI-enabled cybersecurity, clinically validated health AI, regulated enterprise deployments, fintech and identity, and enterprise automation rather than consumer chatbots. Across 2024 to 2025 tracking and research (PitchBook, CB Insights, a16z Signals, McKinsey), the pattern repeats: Infrastructure categories pull capital, regulatory tooling scales, and regulated segments like health AI and compliance fintech keep getting premium attention.

An anecdote makes the stack tangible. Figure AI shows how capital gravitates toward companies that control not just models, but embodied AI platforms where intelligence meets physical labor, safety constraints, and real-world liability. The barrier to entry isn't code; it's capital intensity, and governance.

The implication is simple: "AI investments" aren't one category anymore. The question is where in the stack you invest or build: chips, cloud, foundation models, distribution, data, compliance, or sector workflows. The survival strategy follows: Don't compete with the model; compete with distribution, trust, and domain workflows, and own the customer relationship. Early exits and M&A reinforce the same point: Investors are valuing repeatable revenue and safety credentials over story, and a growing number of companies are finding outcomes through acquihires or mid-tier M&A as enterprise adoption outruns mainstream hype. Early acquisition happened to Touchcast that offered the complementary lesson higher up the stack: an enterprise AI and immersive communications platform where the best outcome came not from novelty, but from distribution, culminating in its 2025 acquisition to expand AI-powered solutions into broader enterprise workflows and interfaces.

The Regulatory Slow Down

We're entering an era of regulatory fragmentation: state-level AI laws in the US alongside major frameworks like the EU AI Act, plus APAC pilots and sandboxed environments. Governance isn't global. It's fracturing, and that complexity can either crush you or become your moat. The investor logic is straightforward: Become compliant in one major region, and you can become the default provider for compliant framework elsewhere because nobody enjoys building compliance twice.

One prevailing theory argues that heavy-handed AI regulation framed as protecting workers from displacement will slow innovation by raising compliance costs, discouraging experimentation, and pushing cutting-edge development to less regulated jurisdictions. Critics counter that while regulation may delay short-term deployment, it shapes where and how AI scales, favoring firms that embed safety, labor transition, and accountability into their business models rather than eliminating progress altogether. There have been attempts to pause or limit regulation, but the broader direction is clear: Governance is accelerating, not disappearing.

China is running a distinct playbook. AI is anchored at the center of national strategy, backed by top-down planning, massive support for R&D, compute, talent, and adoption, and coordinated public-private ecosystems that diffuse AI across industry and public services. In parallel, China is building comprehensive model and service oversight systems, a hybrid of innovation encouragement and societal control. Unlike the West's market-led approach, China's ecosystem is strategically planned, state-enabled, and tied to national objectives in economic transformation, sovereignty, and governance capacity.

Healthcare makes the rule brutally obvious: In regulated sectors, "move fast and break things" becomes "move fast and get banned." Sehat Kahani shows how scaling health services in emerging markets requires clinical oversight, privacy and cultural context. Isaac Health demonstrates the same posture in mental health: Human-in-the-loop care, clinical accountability, and regulatory literacy matter more than model sophistication. The practical advice is boring but undefeated: Build literacy in AI governance—model risk, bias, provenance, documentation—and learn to translate between product, legal, and regulators without starting a small war.

"Internet as a Bank" Is Becoming Real

The financial system is being rebuilt around tokenized settlement assets—stablecoins, deposit tokens, wholesale CBDC pathways—which means money is shifting from "banking product" to "network primitive." Major payment networks and banks are pushing stablecoin settlement in production contexts, which is the tell that this is moving from theory to architecture. In parallel, AI is increasingly paired with programmable money rails in fintech. Capital is flowing toward startups that combine intelligent automation with compliant rails: AI-based credit decisioning, AI-enhanced AML/KYC, and autonomous payment flows are no longer futuristic. They're becoming standard deal narratives.

We're also seeing countries treat crypto rails as financial infrastructure rather than fringe innovation. Jurisdictions including the United Arab Emirates, Singapore, Switzerland, Argentina, and Nigeria are moving toward permissive or proactive frameworks around stablecoins, tokenization, and digital assets, driven either by national strategy or macro necessity.

Two examples make it feel real. Gozem, a West African super-app spanning mobility, logistics, payments, and finance, shows how digital platforms become financial solutions when they integrate payments and identity into a single rail, amplifying inclusion only when governance is embedded. InvestEd highlights how access to capital and financial literacy becomes mediated by algorithms. The question, then, isn't whether AI can make decisions, but who those decisions serve and how transparent they are. The broader implication is unavoidable: AI plus programmable money enables automated commerce at scale—KYC, fraud controls, settlement, lending decisions, payouts—expanding inclusion or scaling extraction depending on incentives.

Legacy giants like Telefónica and other global telcos are emerging as quiet but critical players in this AI-finance stack because they sit on real-world identity signals at population scale—SIM ownership, device integrity, location patterns, network behavior—that few others can legally or reliably access. Through network APIs (standardized, permissioned ways for software to request network-level verification), telcos can expose identity verification, SIM-swap detection, fraud prevention, and KYC assurance without sharing raw personal data. This turns telcos into intermediaries, not just connectivity providers. Telefónica's partnership with Chainlink extends this logic by connecting telecom identity and network signals into blockchain and smart-contract ecosystems for verification use cases in antifraud, payments, and digital identity. As AI systems automate financial decisions and transactions, telco APIs provide the missing layer of grounded verification that anchors digital intelligence to physical infrastructure.

The advice stays pragmatic: Understand stablecoins, tokenized cash, and embedded finance, and the identity and compliance gates that will decide who gets access.

Labor Disruption

The labor debate is shifting from jobs lost to income design. AI compresses labor value across many white-collar workflows and may collapse traditional labor-cost advantages over a five-to-ten-year horizon. Emerging markets will need to move from "cheap labor" to intelligent architecture, and in APAC the opportunity lies in federation rather than domination. The region is structurally diverse—languages, regulations, market sizes, digital maturity—so a single dominant platform is unlikely. Federation—shared standards and APIs connecting AI systems, identity frameworks, and payment rails—offers scale without sacrificing sovereignty, turning fragmentation into a moat.

As inequality becomes politically unstable, societies will experiment with UBI, wage subsidies, portable benefits, and automation-linked redistribution. But UBI is not a magic fix: Distribution, power concentration, and dignity still matter. Embodied AI makes this unavoidable. When robots enter warehouses, factories, and care environments, the issue isn't replacement; it's who captures productivity gains and how society redistributes value. Competition changes too: Embodied AI collapses wage arbitrage and replaces it with capital cost, energy cost, and systems integration. Countries that pair robotics with workforce transition, infrastructure investment, and fair value redistribution will outperform those treating it as pure cost-cutting.

Universal pay would reshape society far beyond economics, creating demographic and psychological fault lines. Younger cohorts, creators, and digitally fluent generalists may experience it as optionality: freedom to retrain, build, care, create, Others may experience it as symbolic redundancy: compensation for displacement rather than recognition for participation. Without pathways to purpose, abundant leisure can slide into disengagement and addiction, amplified by algorithmic entertainment. The deeper risk isn't fiscal sustainability; it's meaning at scale, whether societies build institutions that preserve contribution, dignity, and agency.

The real risk of universal pay isn't economic collapse. It's meaning collapse. Income guarantees can stabilize consumption, but they don't automatically preserve dignity, agency, or a sense of contribution. At scale, inclusion has to be social as well as financial. The institutions that matter most in an AI-augmented economy won't just distribute money; they'll allocate roles, voice, and visible contribution. Think credentialing systems that reward learning and care work, not just output; reputation layers that make non-market contributions legible (mentorship, community building, stewardship); portable guilds or professional cooperatives that let people "belong" outside a single employer; and civic-tech platforms that turn participation—local governance, peer review, public goods—into recognized status rather than unpaid invisibility. Without these institutions, universal pay risks creating a class of economically supported but socially sidelined people, where agency erodes and purpose is outsourced to entertainment algorithms. With them, income security becomes a foundation for experimentation, lifelong contribution, and plural definitions of value. Inclusion, in other words, isn't just about who gets paid. It's about who still gets to matter.

Platforms like Bookr underscore the individual-level shift. As creators and service providers rely on digital rails for income, AI can either extract value invisibly or rebalance bargaining power through transparency and automation. The survival strategy holds at both company and personal levels: Develop cross-domain fluency (AI + regulation, finance, health), build optionality through multiple income streams and portable reputation, and treat your career like a portfolio because it already is.

AI Disruption: Surviving or Thriving? Who Wins, Who Loses, and Why This Is a Choice

The winners will be those who own interfaces, data standards, and relationships; the losers, those reduced to invisible inputs. This outcome isn't inevitable. It's governed by policy, capital allocation, and collective action. The most resilient individuals will seek optionality: portable skills, diversified income, and exposure to AI-enabled platforms rather than dependence on single employers. The "AI shift" won't be won by the loudest demo. It'll be won by companies that solve governance and auditability, cross-border compliance, secure identity and fraud controls, and fair value exchange for data and work, then stitch those solutions into the new money rails. This is where economy, environmental, social, and governance (ESG) becomes practical, not performative. This is where ESG stops being branding and starts being underwriting.

In an AI-driven ESG metrics become operational rather than aspirational. The environmental impact shows up in compute efficiency, energy sourcing, and hardware lifecycle. Metrics like energy per training run, inference efficiency per transaction, and exposure to carbon-intensive data centers increasingly affect margins and regulatory exposure. In terms of social dimension, labor transition becomes an investable signal: workforce reskilling spend, percentage of roles augmented rather than automated, human-in-the-loop coverage in high-risk decisions, and portability of benefits or credentials all map directly to long-term productivity and political stability risk. On the governance side, auditability, model documentation, evaluation coverage, and incident response are no longer "best practices" but measurable risk controls tracked through model cards, audit trails, red-team frequency, and mean time to detect and remediate failures.

For institutional investors, these aren't soft values; they're leading indicators of durability. Companies that can evidence responsible AI deployment across these ESG dimensions price lower regulatory risk, lower headline risk, and lower long-term cost of capital. In that sense, ESG in the AI era isn't about virtue. It's about survival.

In regulated sectors like finance, health, and telco, trust—not compute—is the bottleneck. Privacy and compliance are no longer merely constraints; they're moats. Concept Bio sits at that boundary, applying reproductive biotech approaches where adult somatic cells can be reprogrammed into gamete-like cells, an area that demands ethics, consent, and governance as primary design requirements, not a footer in the terms and conditions.

Across these investments and experiences, one pattern repeats: AI doesn't create value on its own. It amplifies the systems it's embedded in. Companies that treat AI as a feature chase speed; companies that treat AI as infrastructure build resilience. The next generation of winners won't be the loudest or fastest but the ones who understand that trust is the scarcest resource in the AI economy.

About the Author

Alex Szomora is an investor and go-to-market strategist at the intersection of AI, fintech, telecommunications, and digital infrastructure. A former managing director in cloud consulting, she now supports technology and telco organizations with complex program delivery, technology standardization, and the launch of regulated, market-ready products at scale. Her work focuses on interoperable systems, ecosystem collaboration, and institutional trust, especially where AI, identity, payments, and governance intersect.

She is also an active angel investor and limited partner, investing across AI, robotics, healthtech, fintech, and emerging-market platforms, and is an LP at 1982 Ventures and Beyond Ventures. Across both investing and execution, her focus is on durable value creation, financial inclusion, and supporting underrepresented founders in fintech, healthtech, femtech, and longevity.

Email: szomoraa@vuvpfund.com

Website: www.eon-protocol.com

LinkedIn: https://www.linkedin.com/in/szomoraa

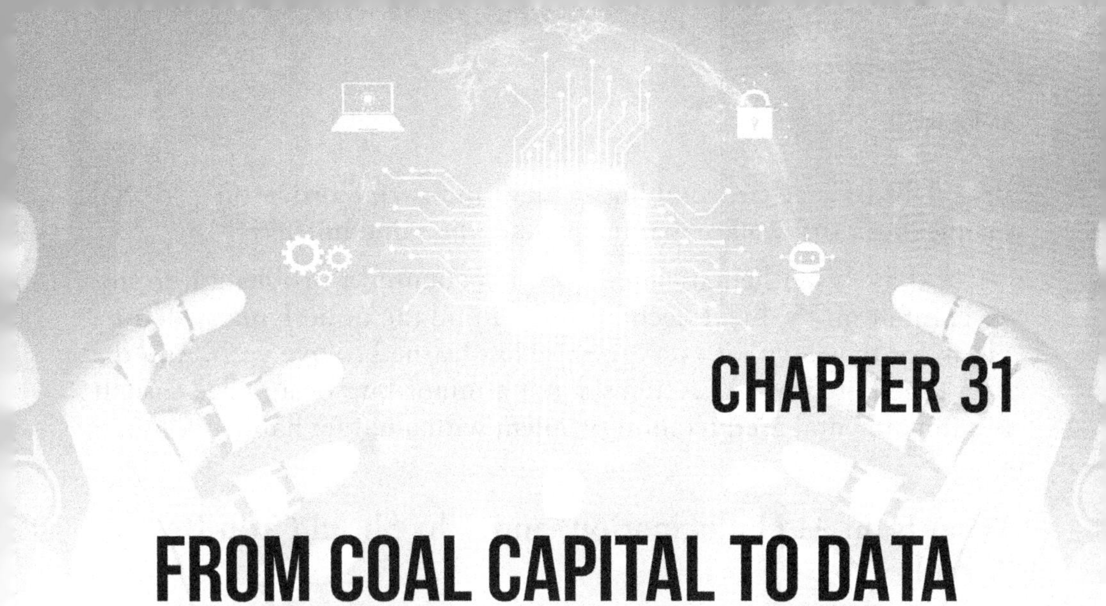

CHAPTER 31

FROM COAL CAPITAL TO DATA CAPITAL: AI HALLUCINATIONS AND DIRE CONSEQUENCES

By Hatice Tavli
Applied Mathematician & AI Inventor
Essen, Germany

All models are wrong, but some are useful.

—George E. P. Box

For 150 years, Essen was the coal capital of the world. Deep beneath the Ruhr Valley, miners extracted the black fuel that powered European industry, heated homes across the continent, and built fortunes that shaped modern Germany. The engineers of the coal era believed they understood their system completely. They were wrong. What they could not see was that their system had a fundamental flaw embedded in its architecture: The resource was finite. The environmental cost was compounding. The collapse, when it came, was devastating and slow enough that no single moment felt like the crisis it was.

Today, we are building a new industrial order on artificial intelligence. And we are making exactly the same mistake.

I am a mathematician. When I encounter a problem, I do not accept it at face value. I decompose it. I find the underlying structure. I look for the flaw that everyone else has learned to live with. And the flaw I see in today's AI systems is not a minor bug or an edge case. It is a fundamental architectural problem with a name: hallucination.

What Is an AI Hallucination, and Why Should You Be Afraid?

A hallucination, in the context of artificial intelligence, is not a metaphor borrowed carelessly from psychology. It is a precise technical term describing a specific failure mode: When an AI system produces a confident, coherent output that has no basis in reality. The system is not lying. It does not know it is wrong. It is executing its programming with perfect fidelity, and that programming is leading it off a cliff.

To understand why this happens, you need to understand what a modern neural network actually does. It does not reason. It does not understand. It recognizes patterns in data and extends those patterns into new situations. A medical AI trained on one million chest X-rays learns to identify patterns associated with pneumonia, cancer, and cardiac disease. When it sees a new X-ray, it asks, "Which pattern does this most resemble?" And it answers with extraordinary confidence.

The danger is that AI has no mechanism to know when it is encountering something outside its experience. When the new X-ray comes from a patient with an unusual condition, different equipment, or a rare anatomical variant—something the training data did not contain—the system does not say, "I don't know." It says, "Eighty-seven percent confident: pneumonia." That confident answer, delivered without hesitation, without doubt, is a hallucination. And in a hospital, a hallucination can kill.

The Anatomy of a Hallucination

Understanding hallucinations requires looking inside the mathematical structure of a neural network with the precision of an anatomist. What is actually happening when an AI system generates a false but confident output?

Modern AI systems are trained by minimizing a loss function across millions of examples. Through this process, the network learns to map inputs to outputs with high accuracy on its training data. But this training objective contains a hidden assumption: that the data the system will encounter in the real world will resemble the data it was trained on. In mathematics, we call this the "IID assumption": independently and identically distributed. In the real world, this assumption is almost never fully satisfied.

The anatomy of a hallucination has three distinct components. First, there is distributional shift: The system encounters an input that differs from its training data in ways that are invisible to the model but critical in the real world. A surgical AI trained on 100,000 standard cardiac procedures encounters a patient with atypical anatomy. A water treatment AI trained on 20 years of historical data encounters an unprecedented contamination event. The input has moved into a region where the model's pattern-matching is unreliable, but the model does not know this.

Second, there is overconfidence: The system assigns high probability to an incorrect output. This is a mathematical property of how neural networks are trained. Standard training procedures do not explicitly reward a network for knowing the limits of its own knowledge. The network learned to produce confident outputs because confident outputs were rewarded during training. It has no internal mechanism to say, "I am outside my area of competence."

Third, and most dangerously, there is silent failure: The system does not signal that anything is wrong. The output looks normal. The confidence score is high. There is no alarm, no warning, no flag raised for a human being to notice. The hallucination passes undetected until the consequences become visible in the physical world, often irreversibly.

This is the anatomy of the problem. Any serious solution must attack all three components simultaneously.

The Mathematics of Protection: Inverse Problems, Uncertainty, Quantum Physics, and Dynamical Systems

In mathematics, when we encounter a problem that is ill-posed—a problem where small changes in inputs lead to large, unpredictable changes in outputs—we do not accept the ill-posedness. We regularize. We add constraints. We use tools specifically designed for unstable, underdetermined problems. This field of mathematics is called inverse problems, and it offers a precise framework for understanding what hallucination protection really means.

An AI system that hallucinates is solving an inverse problem badly. Given sensor data—the observation—the system must infer the state of the world—the hidden reality. This inference is inherently ill-posed: Many different world states could produce the same observed data. Without proper mathematical regularization, without prior knowledge, physical constraints, and explicit uncertainty quantification, the system will collapse to a confident-sounding but incorrect answer. The inverse problem framework tells us exactly what is missing: structure, constraints, and honest uncertainty.

Uncertainty quantification (UQ) is the cornerstone of the solution. Rather than producing a single output—a point estimate—a properly designed system produces a distribution over possible outputs, explicitly representing what it knows and what it does not. This changes the architecture of decision-making entirely. Instead of "Eighty-seven percent confident: proceed," the system communicates, "I am confident within these bounds; outside these bounds, I require human judgment." The confidence interval is not a sign of weakness. It is a sign of mathematical honesty. A system that knows what it does not know is a safe system. A system that does not know what it does not know is a dangerous one.

Quantum physics offers a deeper insight into why uncertainty is not a flaw but a fundamental feature of reality. At the quantum level, nature itself does not produce definite outcomes. It produces probability distributions. Measurement collapses those distributions into specific observations, but the underlying reality is irreducibly probabilistic. This structure propagates upward through chemistry, biology, and into every complex system we build. When we construct AI systems that produce single-point predictions with false certainty, we are violating the mathematical structure of the world they are supposed to model. Uncertainty quantification, grounded in Bayesian inference and quantum-inspired probabilistic logic, returns AI systems to alignment with the actual structure of reality.

The environments in which AI operates—hospital operating rooms, power grids, urban traffic networks, water treatment facilities—are complex dynamical systems characterized by sensitive dependence on initial conditions, nonlinear feedback loops, and emergent behavior that no single model can fully predict. A small error in a complex dynamical system does not stay small. It grows. It propagates. It triggers cascades that no individual component could foresee. Hallucinations in critical systems are not merely unfortunate; they are catastrophically dangerous precisely because of these dynamics. When an AI system hallucinates inside a complex dynamical environment, the error is amplified by feedback until the consequences are visible and irreversible.

The mathematically rigorous approach is to model the AI system and its environment as a coupled dynamical system and to design protection mechanisms that are sensitive to the system's dynamics, not just its static accuracy on a test dataset. This means continuous real-time monitoring, Bayesian updating as new sensor data arrives, and constraint satisfaction algorithms that enforce physical and safety limits at every decision point.

When the Stakes Are Human Lives

Consider what happens when a hallucination occurs in a cardiac surgery suite. A robotic surgical system—guided by AI—is navigating a patient's thoracic cavity. The AI has learned from thousands of successful cases. But this patient has a rare anatomical variant: a coronary artery that branches in an unusual location not present in the training data. The system's pattern-matching engine confidently identifies it as standard anatomy. The surgical instrument proceeds on that assumption. The artery is damaged. The patient hemorrhages. The surgeon has seconds to respond.

This is not a hypothetical. This is the predictable mathematical consequence of deploying a system without hallucination protection in the domain of human anatomy. Consider the power grid. An AI management system encounters a weather event unlike anything in its training data: a simultaneous collapse of renewable generation and a surge in demand. Its pattern-matching machinery finds the closest historical analogy and extrapolates a grid-balancing strategy. The strategy is wrong in ways the analogy did not reveal. Within minutes, the mismatch propagates. Cascading failures ripple through interconnected systems. Hospitals lose power. Water treatment facilities shut down. The complex dynamical cascade, once started, is not easily stopped.

Or consider an autonomous vehicle on an urban street. A perception system hallucinates the absence of an obstacle and assigns high confidence to a clear path where a child is stepping off the curb. There is no warning. There is no uncertainty flag. The system is confident. The vehicle does not brake.

In each of these scenarios, the failure mode is identical: a system operating outside its training distribution, without uncertainty quantification, without constraint verification, producing a confident output that leads to catastrophic action. The hallucination is not a rare exception. It is the predictable behavior of a system operating as designed in a world more complex than its design accounted for.

From Coal to Data: The Lesson Essen Teaches

Essen's coal industry did not collapse because engineers were careless. It collapsed because the entire ecosystem—economic incentives, technical infrastructure, regulatory frameworks, cultural assumptions—was optimized for one approach, and no one had the incentive to question whether that approach contained a fatal flaw at its foundation.

We are in that same position today with artificial intelligence. The global AI ecosystem is optimized for one approach: Build larger models, deploy faster, accept hallucinations as a statistical artifact. The incentives do not reward slowing down to ask, "What happens when this system hallucinates in a hospital? In a power plant? In an autonomous vehicle carrying a family?"

I have spent 18 years working at the intersection of mathematics, artificial intelligence, and industrial systems. I have watched systems succeed and I have watched systems fail in ways that were, in retrospect, entirely predictable from first principles. And I have come to a conclusion I believe is both mathematically rigorous and morally necessary: AI hallucinations are not an acceptable cost of doing business. They are a systematic architectural failure that will cost human lives at scale unless we respond with the seriousness we apply to any other life-safety engineering problem.

Essen teaches us that industrial transitions are possible. They are difficult. They require courage, investment, and the willingness to say, clearly, that the old way has a flaw that cannot be engineered around, only replaced. The transition from overconfident pattern-matching AI to mathematically rigorous, hallucination-protected, uncertainty-aware intelligent systems is exactly that kind of transition.

In mathematics, we do not accept ill-posed problems as given. We regularize them. We add constraints. We quantify uncertainty. We design solutions that are robust to the unknown, because the unknown is exactly where reality lives. The same standards belong in the AI systems we deploy in hospitals, power grids, operating theatres, and the infrastructure of modern life.

The children growing up in cities like Essen deserve an AI-powered world that is as precise, honest, and reliable as the mathematics that describes it. That world is not a distant ambition. The mathematical tools exist. The engineering frameworks exist. The question is only whether we have the courage and the clarity to build it before the cascades begin.

About the Author

Hatice Tavli is an applied mathematician and AI strategist with 18 years of experience in artificial intelligence, blockchain, and industrial transformation. Rooted in the Ruhr Valley—the region that transformed from coal capital to data capital—she brings both personal and professional perspective to the challenge of building reliable, life-safe AI systems. She is the founder of IVDY™ Group, a German-Turkish deep technology company delivering hallucination-protected AI solutions for mission-critical applications: autonomous surgical robotics, high-performance custom AI semiconductors, and hallucination protection software for critical infrastructure. She also founded IVDY Stiftung, providing STEM research and education for children aged 9 and above. Hatice is a published author and international speaker on trustworthy AI and autonomous systems.

Email: hatice@ivdygroup.de

Website: www.ivdygroup.de

IVDY™ Chip—Custom high-performance AI semiconductors: www.ivdygroup.de/chips

IVDY™ Halo—Hallucination protection software: www.ivdygroup.de/halo

IVDY™ Robot—Autonomous cardiac surgery system: www.ivdy.me/ivdy-robot

CHAPTER 32

ENTERPRISE AI ADOPTION

By John Wei
CTO & SVP, Financial Services
Detroit, Michigan

Corporate data continues to expand at roughly 20% to 30% year over year, with most of that growth coming from unstructured sources: documents, emails, chats, images, audio, and video. Until the advent of generative AI, much of this information could only be leveraged through manual effort or narrow, brittle automation.

Many vital corporate functions—procurement, legal, marketing, and HR—have historically scaled at "human scale." For example, when I led a multi-year transformation program at a large bank to retire legacy data centers and migrate applications to the cloud, our vendor landscape changed dramatically. Instead of managing a relatively small set of key technology vendors (e.g., IBM, Fiserv, and FIS), we suddenly needed to manage a much larger number of technology and services partnerships.

Cloud platforms offer a rich set of modern services to compose digital capabilities and better serve customers. Behind that catalog, however, is a complex web of legal agreements that govern third-party relationships and risk. At one point, I was responsible for risk scoring 40,000 contract documents to assess operational risk for regulators. As you can imagine, doing this purely manually—as we have traditionally done, even with a small army of legal and procurement professionals—would be prohibitively expensive and far too slow for a fast-moving digital environment. Using AI to read and score contract documents for risk turned out to be both effective and defensible. Since then, AI has become even more accurate, scalable, and affordable.

Marketing is another area that generates a massive amount of unstructured data: campaign messages, articles, press releases, brand logos and templates, videos, images, and more. How important is marketing to financial services institutions? I once spoke with the CMO of a large bank who described marketing, alongside retail, commercial, and wealth management, as the fourth business unit. Why? Because at its core, a bank takes deposits, makes loans, and facilitates payments. On the main street of any city, many banks can provide these services. What makes a customer choose one over another? People rarely make that decision by studying a 500-page comparison. More often, they respond to messages and experiences that lead them to believe one bank offers greater value.

Today's digital consumers must be reached across multiple channels, multiple times, through a coherent program with consistent messaging. Doing that manually becomes costly quickly. Meanwhile, prospects are researching and searching in new ways. The rise of AI has made traditional keyboard-based search less effective on its own. As a marketer, you now need to provide genuinely useful knowledge, engage audiences beyond clever taglines, and hold real digital conversations with customers and prospects. And as a CMO, you're expected to deliver these outcomes in the new reality at roughly the same budget, certainly not at two to four times the cost that a traditional, non-AI approach might require.

HR faces a similar challenge. Policies, benefits, and guidance materials continue to grow in volume and complexity. If you run global HR operations, once accounting for data privacy rules and country-specific mandates, you can easily end up with a massive body of information that few people read, but that is critical for creating an equitable, fair, and consistent culture. HR professionals want to spend more time supporting colleagues and addressing nuanced change-management challenges, and less time on routine work. Thankfully, AI can handle many of those routine tasks well, such as drafting job postings and supporting resume-to-role matching, freeing HR teams to focus on higher-value work.

Once you see the mountain of unstructured and growing data in the enterprise environment behind the firewall, which AI can help process, it becomes clear that AI is less a threat to replace humans and more a welcome assistant, enabling people to focus on more interesting, non-routine, and higher-value work.

Not all AI is created equal. Deploying AI effectively today requires two things: the skill to match an AI approach to a specific purpose, and the agility to move across build-partner-buy options as the technology landscape continues to evolve and clear ecosystem leaders emerge. Just as an enterprise relies on a mix of specialists, broadly trained hires (from business or engineering programs), and contractors or consultants for short-term needs, enterprise AI solutions similarly fall into three broad categories: domain-specialized AI solutions, general-purpose AI platforms, and homegrown/open-source models.

Domains-Specialized AI Solutions

Domain-specialized AI solutions are typically built on top of general-purpose large language models (LLMs). The vendor invests heavily in assembling domain-specific training datasets, tuning the model for accuracy, mitigating hallucination risk, and delivering user experiences that fit real workflows in that domain. Leah, Anthropic Claude AI Legal Plugin and Harvey are good examples in the legal and contract space.

Because they are optimized for a narrow context, domain-specialized solutions often deliver higher out-of-the-box performance, frequently 80%-plus, and make it easier to configure human-in-the-loop review and approval processes.

General-Purpose AI Platforms

ChatGPT, Microsoft Copilot (including Copilot Studio, the Copilot SDK), and Azure AI services are examples of general-purpose AI platforms. These offerings are typically cost-effective, universally accessible, and backed by mature enterprise security and governance capabilities.

Their trade-off is that they can be less accurate on highly specialized tasks (for example, detailed legal review) unless they are carefully configured and paired with strong controls. That said, when you examine the day-to-day work of many corporate attorneys, only a small portion requires the specialized training acquired through law school. A large share is administrative and general-purpose, such as comparing and merging documents, preparing filings, summarizing communications, or transcribing depositions, where general-purpose AI is often "good enough" and materially cheaper than a domain-specialized solution.

Homegrown and Open-Source Models

Finally, there are homegrown solutions built in-house, often leveraging open-source models. Not every business process is common enough to support a sophisticated commercial software market. For example, our team supports a service line that helps clients polish marketing collateral while adhering to a commercial bank's brand identity. We developed an in-house AI solution that reviews slide content and recommends alternative design concepts. A simple roadmap table, for instance, can often be made more compelling through a wave or pyramid visualization.

In cases like this, developing in-house can be the most practical path. As general-purpose LLMs have become more capable, we have also migrated certain features to general-purpose platforms through API calls. With cloud infrastructure and modern DevOps pipelines, homegrown solutions can be assembled quickly. However, organizations should not underestimate the ongoing costs: maintenance, monitoring, security hardening, and the effort required to keep pace with model and data drift, especially as production deployments expose the solution to larger, noisier, and more varied datasets.

As we evaluate fit-for-purpose AI, it is essential to recognize the distinct trade-offs across cost, speed, differentiation, control, and risk. Understanding these strengths and limitations is foundational to shaping a clear build-buy-partner technology acquisition strategy.

AI adoption is a journey and requires an architected platform approach focused on solving the hard enterprise AI problems. Agentic AI marks a meaningful shift in enterprise AI: moving from answering questions to completing real work, under human supervision. And unlike prior technology waves that delivered capabilities teams could "turn on," AI agents deliver outcomes. That changes the adoption curve: Once agents show tangible value at the task level, they proliferate quickly.

In our own journey developing and adopting AI, the path from a handful of pilot agents to nearly 2,000 AI agent instances, spanning more than 40 agent types, happened fast. Among my peers (CTOs and CIOs, especially in large enterprises), the same pattern is emerging. AI agent sprawl is real.

As deployment scales, the operating environment becomes more complex, often faster than most organizations expect. If that complexity isn't deliberately managed, it turns into hidden cost: duplicated solutions, inconsistent controls, fragmented data access, and a growing maintenance burden that ultimately diminishes the value agents are meant to deliver.

That's why agentic AI needs an architected transformation approach, not random acts of innovation that generate isolated short-term wins but lack the organizational power to scale.

Why the Platform Matters

When you take an architected approach, the goal is a target architecture, a platform that provides a consistent set of design patterns for how AI agents are composed, governed, integrated, and maintained.

The platform approach reduces complexity, enables reuse, and improves cost and operational manageability. Each AI agent can deliver unique functional value, but the underlying plumbing (identity, security, orchestration, logging, integration patterns, evaluation, policy enforcement, and lifecycle management) should be consistent and reusable.

We've seen this movie before in digital and cloud transformations:

- Salesforce is a platform that standardizes core CRM capabilities while enabling differentiation through configuration and ecosystem extensions.
- In financial services, platforms such as nCino build specialized processes on top of broader Salesforce platform capabilities.
- Cloud providers offer layers of infrastructure (e.g., firewall, load balancing, server, storage), platform services (e.g., database-as-a-service or GPU-as-a-service), and software services (e.g., identity management, encryption) that enterprises compose into their own managed environments.

The same pattern is taking root with AI: as enterprises scale adoption, the platform becomes the lever for controlling sprawl and sustaining value.

The Three Hard Problems the Platform Must Solve

A scalable agentic AI platform must consistently address three enterprise-grade challenges:

1. Siloed enterprise technology landscapes
2. Consistent, correct, and complete (3C) data
3. Change management beyond technology adoption

Challenge 1: Breaking Down Technology Silos

AI agents can't thrive in isolation. At scale, agents must evolve into multi-agent capabilities with simpler integration patterns and a more unified operating environment.

Even if a platform vendor offers best-in-class point solutions (for example, contract risk scoring), every enterprise organizes work differently driven by operating model, governance, talent mix, and business priorities. Over time, as AI becomes more accurate and capable, it will also reshape how workflows are designed: where humans oversee, where quality gates sit, and how tasks route across teams and skill levels.

That's why the ability to rewire workflows in low-code/no-code ways without breaking controls is critical to the long-term sustainability of any agent ecosystem.

Here's what this looks like in practice: As an early enterprise-scale adopter of Leah GenAI solution since early 2024, we've deployed hundreds of Leah agent instances and built software assets that integrate Leah agents (including via API calls) to support scaled alternative legal services provider (ALSP) operations. With Leah's announcement of the AgenticOS platform, and its roadmap, we plan to migrate those custom solutions onto the Leah AgenticOS platform.

The strategic benefit is straightforward: Migrating custom code helps us reduce our custom software footprint, reduce ongoing IT maintenance scope, and adopt platform innovations faster, supported by the economics of a software platform that can invest at scale.

In a similar way, we view Microsoft Copilot as a platform foundation. Across Copilot in Teams, Copilot Studio, and SDK-based development, our AI COE has built non-trivial productivity agents tailored to our ALSP environment. In lab A/B time-series testing, we have observed productivity improvements as high as 90%. In production, after accounting for learning curves and data variability, we see closer to ~30% productivity improvement today. A stable platform backbone lets us benefit from ongoing vendor innovation while we focus our energy on the "top layer" of functional refinement and operating model fit.

Challenge 2: Solving the Data Challenge

For legal and regulatory work especially, data has historically been distributed and fragmented, managed case-by-case across corporations, law firms, and providers. As AI adoption scales, we see the trajectory moving toward cloud-native platforms with built-in data governance, helping organizations mature from distributed models to federated models and over time toward more centralized governance where it makes sense.

This is why we're investing in strategic partnerships with Exterro AI, Everlaw, and Leah, with more to come. What these partners share is cloud-native data governance capability that supports a "store once, use many times" approach, giving clients a path to mature their governance without forcing a big-bang conversion.

Similarly, we're strong proponents of the Microsoft Graph approach, which provides a unified API layer across Microsoft 365 data and intelligence, operating inside an enterprise identity and security envelope.

Challenge 3: Driving True Change Management

As established software vendors embed AI into their offerings, the question becomes less "Does it work?" and more "Will it stick?" Sustainable adoption depends on more than adding new technology. It requires new ways of working, new controls, and new accountability.

Before entering partnerships, we validate alignment on the underlying premise: Successful AI adoption is change-management first, technology second.

We're not tied to legacy revenue streams or older channel models. We act as a change agent, partnering with clients to design AI-assisted joint operating models. Internally, we describe this mindset as "disrupt without disruptions" or "innovate with intention."

In our view, a strong platform also needs the commercial flexibility to support different adoption paths, whether that's outcome-based pricing, a direct platform insourcing model, or other structures that best align with a client's mission, compliance posture, and value expectations.

Agentic AI will scale faster than most organizations expect, and with that scale comes sprawl, complexity, and cost unless there is an intentional architecture behind it. The strategic shift is clear: Winning with agentic AI is less about deploying individual agents and more about building (or adopting) a platform that standardizes how agents are composed, governed, and evolved.

A durable agentic AI platform needs to solve three enterprise-grade challenges:

- Silos: Enable multi-agent, cross-system workflows and flexible orchestration that matches the enterprise operating model.
- Data: Provide cloud-native governance that improves consistency, completeness, and reuse of data over time.
- Change: Support adoption through new operating models, controls, and commercial flexibility that aligns outcomes with compliance and mission delivery.

The enterprises that capture compounding returns from agentic AI will be the ones that treat it as a platform-led transformation, not a collection of disconnected pilots.

Conclusion

In the end, the true north of AI adoption is value creation and what it takes to help humans thrive. AI will continue to improve and surpass capabilities and benefits that we can envision today, which is exactly why enterprises should be prudent optimists: Move decisively to capture benefits, while doing the hard work to implement risk-mitigation controls and durable guardrails. With a platform-led approach that tames sprawl, strengthens data governance, and makes change stick, organizations can turn agentic AI into compounding advantage. All in all, the future is bright with AI adoption behind firewalls in enterprises.

About the Author

John Wei is a technology executive with deep expertise in AI/ML and generative AI, cloud, cybersecurity, and compliance, known for leading large-scale digital transformations in regulated industries. He is EVP and CTO at Integreon, where he oversees technology and product engineering leadership and has helped drive AI- and GenAI-enabled solutions that earned major industry awards, including Legalweek 2025 Innovation and a 2025 AI Breakthrough Award. Previously, he served as CTO/SVP at Comerica Bank and CTO/managing director at Emergent Holdings. Earlier roles at DXC and HP included executive client partnerships and global solution delivery, with recognition as a distinguished technologist.

Email: jwei2000@gmail.com

LinkedIn: https://www.linkedin.com/in/john-wei-193a62

CHAPTER 33

THE ARCHITECTURE OF VALUE: SOLVING BUSINESS OBJECTIVES THROUGH AI INNOVATION

By Sam Wong
Data & AI Builder, Innovator, Speaker and Instructor
Toronto, Canada

Innovation distinguishes between a leader and a follower.

—Steve Jobs

The spark that ignited our AI journey didn't come from a boardroom strategy session or a consulting deck. It happened 30,000 feet in the air.

Our CEO and CIO were flying between international offices when the conversation turned to the generative AI boom sweeping headlines. OpenAI's ChatGPT had just gone viral, and everyone from tech analysts to board members were talking about it. Mid-conversation, the CIO messaged me from the plane: "What's our response to this AI moment?"

Unbeknownst to them, I had already been developing a proposal. I had been sketching out the idea of an internal AI incubator, a structured initiative to bring high-impact use cases to life through rapid experimentation, stakeholder collaboration, and responsible deployment. That flight provided the perfect moment to pitch it. I outlined the concept and vision while they were still airborne. By the time they landed, we had alignment. Within days, I was cleared to begin.

This wasn't about jumping on a trend. It was about ensuring our company, one of the largest global beverage brands, was prepared to embrace AI as a long-term business capability. Not because it was cool, but because it could make us smarter, faster, and more connected to our customers and teams.

Reframing the AI Question

One of the most important early mindset shifts was redefining the central question. Most companies start with: "What can we do with AI?" It's a tempting question, especially with so many shiny tools available. But that leads to chasing features instead of solving problems.

Instead, we asked: "What business challenge, if solved, would create the most value?" That question reframes the discussion. It pulls AI out of the lab and into the core of the business.

This clarity helped focus our early efforts. We didn't need hundreds of use cases. We needed one strong one to show what was possible. And we needed business sponsorship, not just technical talent, to make it real.

The Sales AI Assistant: From Concept to Pilot

After receiving support from our CEO and CIO, I approached a progressive business unit president who agreed to partner as the first sponsor. Working closely with sales and marketing leaders in that business unit, we began by co-hosting ideation sessions to uncover pain points.

Three opportunities surfaced quickly: 1) Sales reps were struggling to personalize pitches at scale. 2) There were too many product options for reps to navigate efficiently. 3) Marketing was overwhelmed with content research and synthesis demands.

We decided to pursue a sales AI assistant as our first pilot. The concept had two components:

- A product recommender system tailored by account, geography, seasonality, and product performance data
- A pitch generation assistant that pulled structured data (e.g., sales KPIs) and unstructured content (product descriptions, seasonal insights, region-specific messaging) into a draft format reps could customize quickly

The goal wasn't to replace salespeople. It was to augment them. Give them superpowers. Let AI handle the tedious prep work, so they could focus on relationships, storytelling, and closing deals.

Pilot with Purpose

We rolled out the pilot in waves. The first phase included a small group of sales reps from two regions, focused on a few key accounts. We gathered feedback on usability, trust in recommendations, and whether the tool saved time or added noise.

After refining the interface and recommendation logic, we expanded the pilot to a larger cohort across multiple regions. By the third phase, we had a national rollout underway.

We didn't set an artificial timeline. Instead, we measured impact through two lenses: sales productivity (time saved) and revenue generation (influence on rep performance).

One critical factor was our partnership with the VP of sales and their leadership team. Their involvement helped remove blockers, encourage adoption, and validate that the tool aligned with how the business operated, not how we wished it did.

Several lessons emerged during this process:

- Co-creation beats handoff. Reps helped shape the assistant, making it more usable and more likely to stick. Our initial recommender model and our pitch structure changed drastically through multiple iterations and feedback.
- Start with value, not features. The assistant was useful, not flashy.
- AI literacy matters. We hosted short sessions to demystify how the model worked, which boosted trust and engagement.
- Don't skip the change management. We reinforced that this assistant was not to replace their job but give them more capabilities and time back to focus on spending more time with their customers.
- We also treated this like launching a new product, not just deploying software.
- Allow for continued feedback for enhancements.

Scaling Demand: Marketing the Win Internally

After the pilot's success, other teams began asking, "Can we get something like that?" Marketing wanted help with content summarization. Supply chain leaders asked about demand forecasting. Even finance became curious about automated insights or help with invoice automation.

We created demo videos, hosted internal showcase sessions, and invited team members from across the org to see what we had built. These weren't polished TED Talks. They were transparent, honest, and practical walkthroughs of how we went from idea to impact.

This internal storytelling proved to be one of the most important accelerators of adoption and enthusiasm.

How to Build an AI Incubator

Creating an AI incubator within an organization isn't just about launching a few pilots. It's about establishing a repeatable engine for innovation. This section outlines a scalable model any company can adapt. Each stage is designed to support rapid exploration of high-value opportunities, alignment with business strategy, and sustainable delivery of impact.

1. Establish: Set the Strategic Foundation

This stage focuses on setting up the right conditions for success. Organizations need to answer three foundational questions:

- Why are we doing this now? (Strategic intent)
- Who will lead it? (Governance and sponsorship)
- How will we work? (Operating model and team structure)

A strong AI Incubator requires executive sponsorship, ideally from both technology and business leadership. It also requires a dedicated team: a blend of AI/ML experts, data engineers, business SMEs, and project managers.

Tip: Don't treat this like an R&D lab. Set expectations that business impact is the goal. Clarify roles early: Product owners, tech leads, and domain experts all have a seat at the table.

Example: In our own experience, I reported directly to the CIO and worked with a cross-functional lead team that included engineering, data science, and business strategy.

2. Intake: Surface and Shape Opportunities

This stage is about identifying the right problems to solve. It blends structured intake processes with creative exploration. The recommended methods include:

- Run design thinking workshops with business stakeholders.
- Use intake templates to capture business problems, opportunities, data availability, and potential impact.
- Score ideas with t-shirt sizing and a feasibility/impact matrix.

Pitfall to avoid: Don't just ask for "AI ideas." Anchor brainstorming around real business pain points: cost, inefficiency, complexity, missed opportunities.

Example: Our sales AI assistant was prioritized because it tackled two measurable challenges: rep productivity and revenue optimization. We started with a discovery session, mapped the workflow, and rapidly narrowed the scope to two powerful capabilities (recommender and pitch assistant).

3. Build: Prototype, Test, Deliver

Once a use case is prioritized, rapid experimentation begins. Think like a startup: Build fast, fail fast, learn faster. The suggested workflow:

- One- to two-week discovery sprint (validate feasibility, define MVP)
- Two- to four-week build sprint (working prototype)
- Two-week user testing and iteration (feedback-driven changes)

Key principle: Include users in the loop. Let salespeople, marketers, analysts, or operators shape what's being built. This improves adoption and ensures product-market fit.

Example: In our pilot, reps tested the assistant across real accounts and regions. Their feedback drove interface simplification and added context filters.

4. Learn: Evaluate, Decide, Refine

Every pilot should end with structured reflection: What worked? What didn't? Should we scale, rework, or shelve the idea? Tools for learning include:

- A/B testing results
- Productivity or revenue uplift
- User satisfaction or NPS surveys
- Technical performance (model accuracy, latency)

Tip: Create a "pilot scorecard" to document results and decisions. Treat failure as a learning outcome, not a dead end.

Example: We shelved one feature early when reps found it confusing. That allowed us to double down on the recommender model, which had immediate traction.

5. Market: Evangelize and Expand

This stage is about scale. Even great ideas can fail if they're not supported or understood across the org. Suggested practices:

- Record or create AI-generated short demo videos and post them on your intranet.
- Invite teams to internal showcases or AI fairs.
- Create success playbooks, so other teams can reuse your models, templates, or process.

Tip: Internal marketing builds momentum. It turns isolated wins into enterprise-wide movement.

Example: Our assistant was showcased in multiple internal forums. The result? Requests from other departments poured in. Marketing wanted summarizers, finance asked for anomaly detectors, and supply chain wanted predictive alerts.

Closing Note on the Framework

An AI incubator is not about technology experimentation alone. It's about creating a structured path from idea to business impact. The goal is to equip teams to solve problems faster, test smarter, and scale what works.

This framework can flex. Some orgs may run five to 10 concurrent experiments. Others may focus deeply on one or two. The key is consistent rhythm, clear sponsorship, and repeatable ways to define, build, and learn.

Over time, your AI Incubator becomes more than a program. It becomes a muscle, a new way of working that blends creativity, data, and impact.

The Future of AI in the Enterprise: Data Management as a Catalyst

As organizations move beyond AI pilots into scalable, enterprise-wide applications, one truth becomes increasingly clear: AI is only as strong as the data that fuels it. While headlines often focus on the latest AI models or chatbots, the foundational enabler behind meaningful AI impact is something far less flashy: modern, well-governed data management.

The real future of AI In the enterprise doesn't hinge on having the most advanced algorithms. It hinges on whether companies have the data infrastructure, practices, and culture in place to make those algorithms useful, trustworthy, and scalable.

In many organizations, AI initiatives are becoming a catalyst for fixing long-standing data issues. As teams begin experimenting with machine learning or generative AI, they quickly run into operational data challenges: poor quality, fragmentation, duplication, or data that exists only in tribal knowledge. These roadblocks can stop promising AI use cases before they start. Rather than treating this as a failure, leading organizations are recognizing the opportunity: AI is forcing the enterprise to mature its data practices. This trend is sparking renewed investment in several key data management domains, including:

Domain 1: Data Quality and Trust

AI outcomes depend on accurate inputs. A recommender system trained on outdated or incomplete customer data will produce irrelevant suggestions. A forecasting model built on inconsistent SKU definitions across regions will deliver misleading projections. To address this, companies are investing in data quality monitoring and scoring frameworks; automated cleansing and validation rules; and feedback loops between AI outputs and human users to flag anomalies.

In effect, AI initiatives are reframing data quality from an IT hygiene issue to a mission-critical business enabler.

Domain 2: Master and Reference Data Management

Many AI use cases require a consistent view of key entities: customers, products, vendors, stores. But legacy systems often create silos or duplicate records that undermine AI performance. AI programs are, therefore, renewing interest in master data management (MDM) and reference data management (RDM). These efforts help create a single source of truth for critical data domains; cross-functional agreement on definitions and hierarchies; and cleaner integrations across systems and processes.

As AI becomes embedded in workflows like pricing, personalization, or supply planning, the need for unified entity resolution becomes unavoidable.

Domain 3: Metadata and Knowledge Management

Generative AI thrives on context. But most organizations lack structured ways to store, tag, and retrieve the internal knowledge needed for summarization, search, and synthesis. To close this gap, teams are now treating knowledge as a first-class product. This includes building curated internal corpora with access control; enriching documents with metadata to guide retrieval-augmented generation (RAG); and tagging subject matter experts and source confidence

These efforts enhance not just AI performance, but also overall organizational intelligence.

Domain 4: Data Governance and Responsible AI

AI at scale requires clarity on data ownership, access, and accountability. Who can use what data, for what purpose, under what policy? What logs exist to track model decisions? How are sensitive attributes handled in model training? Responsible AI isn't a separate layer. It's built on modern governance frameworks that span both data and models. This includes policy-driven data access; model monitoring and explainability; and risk assessment embedded in AI development workflow.

The future of enterprise AI will reward organizations that treat data as a product and build the operational muscle to manage it with rigor and care. That means forming cross-functional teams with data stewards, product owners, and domain experts who shape data pipelines not just for analytics, but for real-time, intelligent applications.

It also means building for reuse and scale with shared data services, versioned datasets, and lineage tracking that support agility without chaos.

Ultimately, AI success isn't about chasing the next breakthrough model. It's about ensuring that every model has a solid foundation of clean, connected, and context-rich data to stand on.

Organizations that embrace this shift—from model-first to data-first—will not only unlock more value from their AI investments, but also build a durable competitive advantage in an increasingly intelligent world.

Final Reflection

The true power of AI doesn't lie in its algorithms. It lies in its ability to deliver meaningful business outcomes.

Across industries, the rush to "do something with AI" has often outpaced a clear understanding of why. The most successful initiatives don't start with a shiny tool or the latest model. They start with a clearly defined business challenge, a measurable outcome, and a team empowered to experiment, learn, and adapt.

What we've learned is that an AI incubator program is more than a sandbox for pilots. It's a strategic mechanism that connects vision to execution, a disciplined structure to surface high-impact opportunities, co-create with the business, and test new ideas safely, quickly, and effectively. It ensures that AI isn't relegated to isolated innovation labs or pockets of experimentation, but is integrated into the heart of how a company thinks, operates, and grows.

Equally important is a focus on data management. You can't scale AI without trustworthy, well-structured data. In fact, one of the biggest lessons from our journey was realizing how often AI efforts were slowed, not by technical limitations, but by fragmented systems, inconsistent definitions, and a lack of metadata or lineage. Every successful AI project forced us to elevate our standards for data quality, master data governance, and knowledge capture.

In hindsight, our sales AI assistant was as much a data transformation project as it was an AI initiative. It required collaboration across domains to align KPIs, clean inputs, and shape a reliable knowledge base. The technical lift was real, but the cultural lift—getting teams to think differently about data ownership, reuse, and shared accountability—was equally vital.

Looking ahead, the organizations that thrive with AI won't be those with the most advanced models. They'll be the ones that build repeatable, cross-functional systems for innovation. They'll prioritize outcomes over experiments. They'll treat data as a product, not just a byproduct. And they'll empower teams not just to consume AI but to create with it.

Our journey began with a question at 30,000 feet. It taught us that progress isn't about having all the answers. It's about building a culture where asking the right questions leads to bold, iterative action.

Sometimes, the most transformative innovations start not with a grand strategy but with the simple willingness to say, "What if we just tried?"

About the Author

Sam Wong does not just talk about data and AI; he builds the frameworks that make it work. As a seasoned data executive based in Toronto, Sam has led complex data and AI transformations across a wide range of industries, including CPG, retail, and real estate.

Sam's expertise lies in bridging the gap between strategy and execution. He is the architect behind IBM's accelerator catalog for data and AI and the creator of a corporate AI incubator program designed to operationalize AI. From sales and marketing to finance and to supply chain, he has successfully deployed AI solutions that drive tangible ROI.

Currently a member of the AWS CPG executive advisory council, and Ataccama's customer advisory board, Sam is a globally recognized voice in the community. He was named a 2024 Consumer

Goods Technology Visionaries for his work in data and AI. He also shares his insights as a speaker and panelist at numerous events throughout the world. When not leading data and AI transformation, Sam facilitates workshops at TDWI and serves as a guest lecturer at universities across North America.

Email: sam.wong@uwaterloo.ca

LinkedIn: https://www.linkedin.com/in/sam-wong-data-ai/

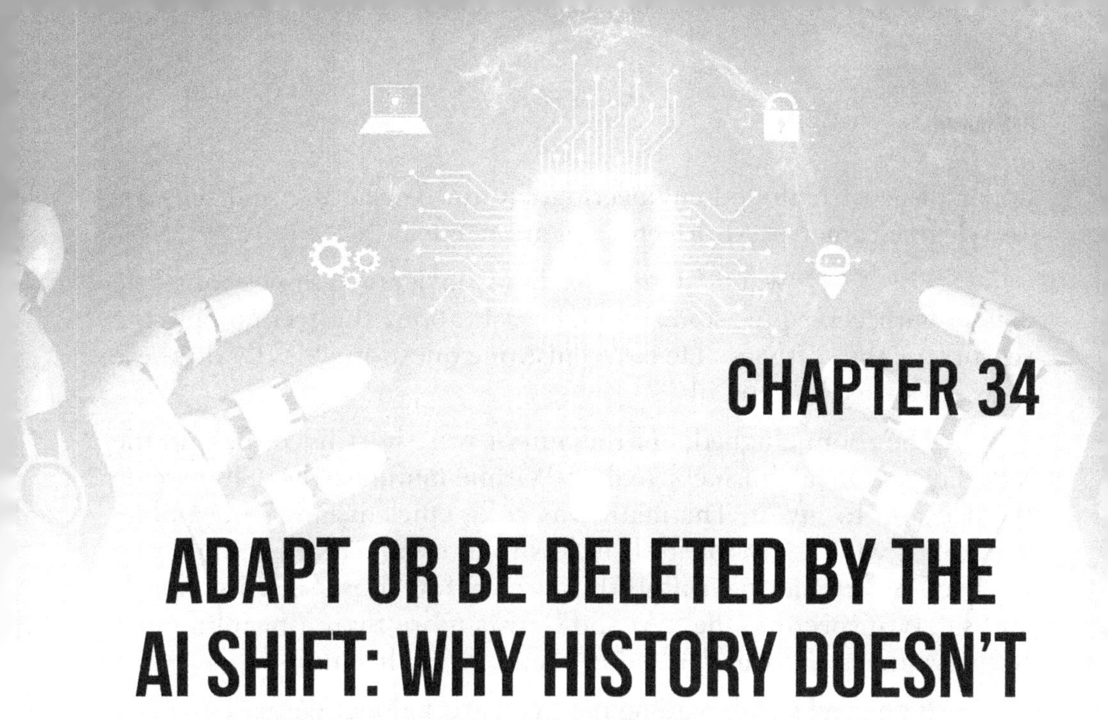

CHAPTER 34

ADAPT OR BE DELETED BY THE AI SHIFT: WHY HISTORY DOESN'T CARE ABOUT YOUR EXCUSES

By Haonan Wu
Data & AI All-Rounder
London, England, United Kingdom

In God we trust. All others must bring data.
—Dr. W. Edwards Deming

I. THE CEO AND THE MECHANICAL HOUND

The ballroom of leadership conference was silent, save for the faint, hydraulic whir of a four-legged machine.

A few years ago, I witnessed a pitch that would have seemed like science fiction just a decade prior. A robotics firm was presenting a““robotic do”” to the CEO of a multi-billion-dollar global manufacturing company. This was”t a toy; it was a million-dollar piece of high-end engineering equipped with thermal sensors, 360-degree cameras, AI-driven navigation, etc.

The CEO watched the machine navigate a mock obstacle course with eerie precision. He did"t ask about the technology, the sensors, or the software. He asked just one question: ""Why do these dogs have legs, but no tails""

The room laughed, but the humor was short-lived. Behind the joke lay a critical business reality. Within minutes, the CEO made the decision to invest. The math was cold, efficient, and undeniable: The robot would pay for itself in less than two years by eliminating the need for human payroll in those roles. For those security guards and safety inspectors, the""AI shif"" was more than a headline in a business journal. It was the end of their current livelihood.

If you feel a chill reading that, you are"t alone. We are currently standing in the shadow of a wave of automation and artificial intelligence that threatens to wash away roles we once thought were""safe"" But before we succumb to the fear that humanity is being""deleted"" we must look backward.

II. HISTORICAL ECHOES: LESSONS FROM THE PAST

The story of the robot dog is not a new story. It is the latest verse in a song that has been playing for centuries.

The Ghost of the Coach Driver

Imagine 1920. If you were a professional coach driver, you were the backbone of urban transport. You understood horses, you knew the city's cobblestone veins, and you held a respected trade. When the""horseless carriag"" began to chug down the streets, you likely viewed it with a mix of mockery and terror. It was loud, it was unreliable, and, most importantly, it was a threat to your famil"s table.

Fast forward: The horses are gone, and the coach driver is a historical footnote. Yet, are we worse off? The automotive revolution did"t just replace the driver; it created the mechanic, the highway engineer, the suburban developer, and the long-haul trucker. It traded the "stabl" for the "superhighway" raising the global standard of living in ways a coach driver could't have imagined.

The Light That Failed

Even the most elite "expert" arn't immune. Just less than 20 years ago, the titans of lighting—GE, Philips, Siemens—employed the world's most brilliant chemists and electronics researchers. These were Nobel-prize-level minds dedicated to the physics of the perfect filament and the chemistry of incandescent gases. They were the masters of their universe.

Then, the LED became commercially viable.

Almost overnight, "lighting" stopped being a field of chemistry and became a field of semiconductors. Those massive research labs and specialized plants became obsolete. The experts who had spent 40 years mastering old science suddenly found their deep expertise was a relic. The "smart" ones didn't mourn the light bulb; they realized that while the medium had changed, the human need for light had not. They pivoted from being chemists to being electronics innovators.

And Many Other Similar Stories

Before the 1950s, "computer" were humans (mostly women) doing manual calculations. Elevator operators: We used to pay people to pull levers to stop elevators level with the floor. The Luddites (textile workers): Often misunderstood as "hating technology" they actually hated that machinery that allowed factory owners to bypass skilled labor.

The Lesson for the AI Era

The common thread through history is simple: Innovation does not ask for permission, and it does not wait for you to be ready. Whether it is a carriage driver in the 1920s, a computer in the 1950s, a chemist in the 2010s, or a security guard in the 2020s, the cycle is the same. But here is the secret that history hides in plain sight: Technology rarely replaces humans; it replaces the tasks. In this chapter, we are going to look past the "robot dog" and the "AI chatbot" to understand the mechanics of these shifts. We will explore how to judge if your role is at risk, how to synthesize the lessons of the past, and, most importantly, how to ensure that when the "AI shift" comes for you, you aren't the one being replaced. You're the one leading the transition.

III. THE REPLACEMENT RISK MATRIX

How do you know if you are the "coach driver" of the 21st century? To determine your risk, we must look at two dimensions of your daily work: process predictability and human complexity.

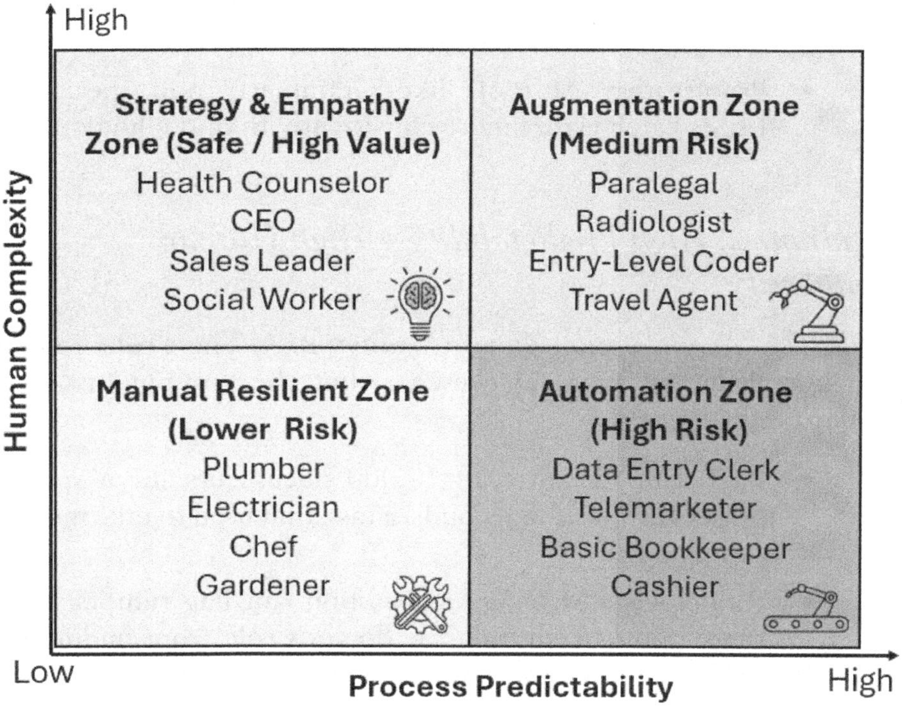

Quadrant 1: High Predictability + Low Human Complexity

Status: The "Automation Zone" (High Risk) These roles are based on structured data and repeatable "if-this-then-that" rules.

- Data entry clerk: Sorting and inputting text is now handled by optical character recognition (OCR) and LLMs with 99% accuracy.
- Retail cashier: Replaced by self-checkout and "just walk out" vision systems.
- Telemarketer: Scripted calls are being replaced by hyper-realistic voice AI that never gets tired or discouraged.

- Basic Bookkeeper: Software now automatically reconciles bank statements and categorizes expenses.
- Proofreader: AI tools like Grammarly and specialized LLMs catch errors more consistently than the human eye.

Quadrant 2: High Predictability + High Human Complexity

Status: The "Augmentation Zone" (Medium Risk) These roles require deep knowledge or physical presence, but the "output" is often standardized.

- Paralegal: AI can scan 10,000 documents for a specific "discovery" fact in seconds, a task that used to take months of human labor.
- Radiologist: AI image recognition can flag tumors faster than a human, shifting the doctor's role from finding the problem to managing the patient's treatment.
- Entry-level coder: AI can write "boilerplate" code (the repetitive parts). The human must shift to being a "system architect."
- Travel agent: Booking a flight is automated; however, planning a complex, multi-country luxury honeymoon still requires a "human touch."
- Translator: Basic manual translation is gone; the new role is "cultural editor," ensuring the spirit of the message survives.

Quadrant 3: Low Predictability + Low Human Complexity

Status: The "Manual Resilient Zone" (Lower Risk) These are hands-on jobs in "unstructured" environments where a robot physically cannot maneuver yet.

- Plumber: Every house is different. Finding a leak behind a specific 50-year-old wall requires "gut feeling" and physical dexterity AI hasn't mastered.
- Electrician: Working with live wires in cramped, non-standard attics is too variable for current robotics.
- Chef (line cook): While a burger-flipping robot exists, a kitchen during a "dinner rush" is a chaotic environment that requires high-speed physical adaptation.
- Gardener/landscaper: Dealing with living, changing organisms in varying weather and terrain is highly unpredictable for machines.

Quadrant 4: Low Predictability + High Human Complexity

Status: The "Strategy and Empathy Zone" (Safe/High Value) These roles rely on what makes us uniquely human: leadership, ethics, and emotional navigation.

- Mental health counsellor: AI can provide a script, but it cannot "hold space" for a grieving person or build the trust needed for deep healing.
- CEO/strategic leader: High-stakes decisions involve politics, ethics, and "betting on a vision" where there is no historical data for AI to study.
- Negotiator/sales leader: Closing a multi-million-dollar deal isn't about the data; it's about reading the "vibe" in the room and building a relationship.

- Social worker: Navigating a complex family crisis requires empathy and moral judgment that cannot be reduced to an algorithm.
- Teacher/mentor: Beyond just "giving information," a teacher inspires and motivates. AI is a textbook; a teacher is a spark.

IV. THREE STRATEGIES FOR SUCCESS

If your job currently sits in the Automation or Augmentation Zone, you aren't at a dead end, you are at a crossroads. To stay relevant, you must change how you provide value.

1. Move "Up the Stack"

In the world of computing, the "lower stack" handles the raw, invisible processing, while the "upper stack" manages the user experience and high-level logic.

- Strategy: Identify the task AI has commoditized and move one level above it. If you are a proofreader, don't just check grammar; become a content strategist who ensures the tone of voice drives sales. If you are an entry-level coder, stop focusing on syntax and start focusing on system architecture.
- Real-world shift: Think of the paralegal. Instead of spending 60 hours a week on document discovery (the lower stack), they must move up the stack to become case analysts, using AI to find patterns that help a lead attorney build a winning narrative.
- Goal: Stop being the "doer" of the task and start being the "architect" of the outcome.

2. Become a "Centaur"

In the 1990s, after Garry Kasparov lost to Deep Blue, he pioneered "Centaur Chess," human-AI teams. These teams consistently outperform both the smartest humans and the fastest supercomputers.

- Strategy: Embrace the machine as your "exoskeleton." If you are a radiologist, don't compete with the AI's ability to spot a lesion. Use the AI to scan 1,000 images in seconds, then use your human expertise to provide the empathy and nuance required to explain the results to a terrified patient.
- Real-world shift: A travel agent shouldn't spend time booking a simple flight. A bot does that better. A "Centaur" travel agent uses AI to curate 50 different itinerary options instantly, then uses their human intuition to pick the one that perfectly matches a client's unspoken preferences.
- Goal: Use the machine to handle 80% of the volume, so you can provide 100% of the value in the final 20% of the work.

3. Solve for "Ambiguity"

AI is a "prediction machine"; it uses the past to guess the future. Therefore, AI is useless in "zero-data" environments where there is no history, only chaos and gray areas.

- Strategy: Lean into roles that require moral judgment, political navigation, or creative invention. AI can write a script for a telemarketer, but it cannot navigate the complex, emotionally charged "gray area" of a social worker trying to keep a family together.
- Real-world shift: While a basic bookkeeper is replaced by automated software, a strategic CFO thrives by solving ambiguity, deciding whether to invest in an unproven new market or how to navigate a sudden global supply chain crisis.

- Goal: Position yourself in the unknown. While AI optimizes the known past, humans must still lead us into an unpredictable future.

V. THE SHIFT—YOU LEAD IT

The CEO I mentioned at the start of this chapter didn't buy a robot dog because he hated his staff; he bought it because he loved efficiency. The "AI Shift" is not an emotional movement; it is a mathematical one.

History shows us that every time a door closes on a manual task, a window opens for a strategic leader. The coach drivers who became fleet managers thrived. The chemists who became semiconductor experts led the world into a new era of LED lighting. The question isn't whether AI will change your job. It's whether you will be the one driving the change or the one being left on the side of the road.

The transition from a "doer" to an "architect" doesn't happen by accident. It requires a deliberate audit of your current value and a roadmap for augmentation.

I help leaders and teams navigate this transition through bespoke "Future-Proof Audits" and "Centaur Mindset" workshops. We don't just look at what AI can do to you; we look at what you can do with AI. The future belongs to those who see it coming. Let's build your future-proof career together.

A note on the making of this chapter: In the spirit of the "Centaur" mindset discussed in this chapter, the author collaborated with Gemini (Google's AI) for content polishing, visualization, and table structuring. This collaboration serves as a real-world example of how human strategic vision and AI execution can combine to move "Up the Stack."

About the Author

Haonan Wu is a global leader in AI and data strategy, currently serving as the group head of data strategy and enterprise governance at Coca-Cola HBC. With a distinguished career spanning McKinsey & Company, Accenture, and Synechron Business Consulting, he has a proven track record of scaling AI from "lab to operation" across diverse sectors and global organizations.

Haonan's perspective is forged by the industrial shifts he describes above. Having begun his career as a data scientist at Philips and Siemens even before the title of data scientist became a household term, he witnessed firsthand the transition from traditional engineering to AI-driven autonomous systems.

Today, he also serves as the headmaster of the Coca-Cola HBC Data & AI Academy. In this role, he leads a strategic transformation program upskilling over 5,000 cross-function business managers of the organization to move "Up the Stack" by evolving into "AI-Savvy Business SMEs." A recognized expert in his field, Haonan is the core author of several research papers on the applications of AI and machine learning.

Email: haonan.wu@hotmail.com

LinkedIn: www.linkedin.com/in/haonanwu

DID YOU ENJOY THIS BOOK?

If you enjoyed reading this book, you can help by suggesting it to someone else you think might like it, and **please leave a positive review** wherever you purchased it. This does a lot in helping others find the book. We thank you in advance for taking a few moments to do this.

THANK YOU

You might also like other Thin Leaf Press titles:

The AI Advantage: Thriving Within Civilization's Next Big Disruption

The AI Revolution: Thriving Within Civilization's Next Big Disruption

The AI Mindset: Thriving Within Civilization's Next Big Disruption

AI: Work Smarter and Live Better Within Civilization's Next Big Disruption

Peak Performance: Mindset Tools for Managers

Peak Performance: Mindset Tools for Sales

Peak Performance: Mindset Tools for Leaders

Peak Performance: Mindset Tools for Business

Peak Performance: Mindset Tools for Entrepreneurs

Peak Performance: Mindset Tools for Athletes

The Successful Mind: Tools to Living a Purposeful, Productive, and Happy Life

The Successful Body: Using Fitness, Nutrition, and Mindset to Live Better

The Successful Spirit: Top Performers Share Secrets to a Winning Mindset

Winning Mindset: Elite Strategies for Peak Performance

Winner's Mindset: Peak Performance Strategies for Success

The Life Coach's Tool Kit, Vol. 1

The Life Coach's Tool Kit, Vol. 2

The Life Coach's Tool Kit, Vol. 3

Ordinary to Extraordinary

The Magical Lightness of Being

Explore.

Made in the USA
Monee, IL
05 November 2026

67911912R00216